MORALITY
VS.
REALITY

HOW THE POWERS OF THE WORLD ARE SHIFTING FROM RELIGION TO MEDIA IN ORDER TO GAIN CONTROL OVER PEOPLE

By

Farooq Tareen

DEDICATION

To all peace-seekers in the world.

CONTENTS

ACKNOWLEDGMENTS

I would like to express my gratitude to my family and friends; they have continued to sharpen my interest with their patience and perseverance. The project could not have reached its conclusion without the help of a few very special people. **Tauseef Ahmed Qureshi,** who initially encouraged and inspired me to put my pen to paper and remained accommodative and supportive throughout. His father was my most admired teacher whose irreprehensible[10] character was a great influence on boosting my self-esteem.

I am deeply thankful to **Sami-Ullah Khan**, from the Institute of Business Administration in Karachi, Pakistan. He managed to read my hand-written scribbled notes and converted them into a typed, draft manuscript.

I owe my gratitude to **Kishan Adoni**, a PhD student at Birmingham University. During his holidays, he reviewed every word of this work. Kishan has the finest combination of scientific knowledge, language, skills and expertise in IT. He is well merited for going through every page of this book.

My feelings of obligation to my nephew **Zubair Ahmed Khan**, who put the finishing touches to this work by re-organising the headers and devising the index. He also identified a specific problem for the readers of the subcontinent and addressed the issue by translating the uncommon and unconventional English words into Urdu. He then compiled and formatted the standalone numbered glossary of the definitions.

Finally, my heart-felt gratitude to **Steve Bolton**, my business partner and close friend for the past 37 years. I always felt at ease discussing all topics of the work with Steve. He was also kind enough to allow me to use his poem 'Peace' that he wrote when he was just a young man. The poem sums up the contention and is a fantastic notion of peace.

PEACE

What is the meaning of the word?
Does it have a place in a troubled world?
What is Peace? A line in the sand
Creating war for a strip of land

What is Peace when countries collide!
Another war with nowhere to hide
What is Peace? Will it ever be real?
When people are starving and have no meal

What is Peace but a simple Word
That can change the pattern of our World
Through friend and foe we must find
The key to the future of Mankind

When will the corrupt be taken to task?
This is the question we must ask
Before it is too late will peace prevail
To tell its proud tale

When peace drew Palestinians and Jews
To live together and see it through
In this world of Political Storm
Let's stand back and see reform

Religions attempt to justify their positions
But doctrines based on truth or superstitions
So what is the basis of these religions?
That create so many divisions

The hypocrisy that is told
To bring people into fold

The seeds are sewn for extreme beliefs to grow
With each of the sects in the know
What started as Peace to mankind
Is now very hard to find

Mankind has yet to witness its finest hour
When Utopia is not control or power
When there is respect and help for fellow mankind
Let's hope it is not too late to find

Peace will always be a compromise
Not consumed by hate or lies

People will have to give and take
If they want to have a stake

So let's skip the rhyme
We don't have the time
The rhetoric in this poem has foundations in idealism
But let's hope that this turns into realism

Utopia this verse may be
But we will have to find the key
To secure the key of tomorrow
If we are to avoid more sorrow

Let the world's leaders proclaim
Let's put hate to shame
Give Peace a chance before it is too late
And make this world a better place

Wishes of peace are lost in the mist
Time to force belief with an iron fist
Let the world leaders proclaim
Let's put hate to shame

Peace will always be a compromise
Not consumed with hate or lies

People will have to give and take
If they want to have a stake

Give Peace a chance before it is too late
AND make the world a better place
For mankind to have its finest hour
When utopia is not controlled by power

Stephen Leslie Bolton

PROLOGUE

In 1947, partition of the subcontinent witnessed the outbreak of brutal atrocities against humanity. For centuries, a secular consensus had existed amongst the people of the subcontinent. Communities co-existed happily before the political leaders, using religion as a tool, began to provoke communities. People killed each other in the name of religion; Hindus, Sikhs and Muslims confronted each other with unimaginable violence and committed savage brutalities. I was only a young child when my family was forced to flee India into Pakistan. Vivid memories of the horrendous journey and the scenes of carnage, with images of people being killed, have been engraved in my subconscious. Those painful, traumatic and unpleasant events were difficult to cope with; thus, the memories of those horrific images were pushed out of the conscious into the hidden reservoir of the subconscious mind.

1948 saw the catastrophic expulsion of Palestinians from their Homeland. The policy was brought about to reward the Zionist ideologists in the aftermath of the holocaust. The Muslim world was forced to react to this involuntary migration. As such, it gave rise to organisations such as the 'Palestine Liberation Organisation' (PLO) in Palestine, and the Muslim brotherhood in Egypt. These political entities were contrived[1] as a means of creating a counterbalance against the unjust policies of the west. One can theorise that 'Islamic Activism' was a direct consequence of this, ranging from educational reforms to 'Jihad'.

Resentment towards western policies intensified in Pakistan as well. Various political and religious groups jockeyed for power, taking full advantage of the unstable landscape arising from Palestine's upheaval. Among the various political and religious groups, Jamaat-E-Islami emerged as a powerful and popular organisation. Its objective was to mitigate western values through critical analysis and to introduce newer dimensions of 'the moral excellence', based on the pattern of the earliest 'Caliphate and Sharia laws'.

As I grew up, I was influenced by social preconceptions[2], synthesised[3] by religion, accessible literature and the popular scientific theories. In truth, I struggled to find the 'correct' world-view! I was fond of Urdu literature and the popular works of contemporary writers such as: Saadat Hasan Manto, Rajinder Singh Bedi, Krishan Chandar, Ismat Chughtai and Quratul-ain Haider impressed me. These scholars made me realise that if all humanity shares in the common concepts of needs, comfort and enjoyment, then why can they not share in the physical, intellectual and spiritual problems of the human existence?

History is a continuous record of warfare initiated in some part by religious, sectarian, racial and political causes. Why has each religion and its teachings boasted of its own superiority and excellence, consequently damaging the validity[4] of all others? If all religions were founded upon the divine love for humanity, then why is there discord, enmity[5] and blind limitations amongst various beliefs? I want religion to be the divine remedy for human antagonism[6] and disharmony. We appear to have made the 'remedy as a cause of disease', perhaps it would be better without the remedy! I wish all beliefs, sects and denominations to become one, considering all religions are saying essentially the same things, from a different perspective. One can lay the burden of responsibility on the representatives of different faiths. It is in their power to realise and act upon these concepts of togetherness and multi-religious harmony. I believe that splitting people into hundreds of denominations, differing only in trivial matters, is wrong. We ought to rise above the boundaries of class, creed[213], colour and caste. Religion can be used to express ideas and concepts, and thus, achieve salvations by sharing the common human values.

Growing up in Pakistan in a strict Muslim family, I understood that a large faction[182] of Muslim scholars are at odds with scientific theories such as cosmology and the biological evolution[118] of humans. Religious scholars did not like these scientific theories within the context of Islam. Prevailing theories, including 'The Big Bang' and

'Darwin's Model of Evolution' were among taboos. These scientific models challenged longstanding religious beliefs, and debate on such issues was strictly discouraged. After all, for thousands of years, philosophers and prophets had established the answers about the creation of the universe; either through analytical reasoning or through religious divinity[105] of 'Prophetic Means'. All 'biblical' religions preach that God created the universe in six days. However, in the last couple of centuries, scientists have calculated the 'creation myth' through mathematical observations.

In 1964, at the age of twenty, I migrated to England to seek better economic opportunities and hoping to provide a secure future for my family back in Pakistan. Work in England was plentiful, and the opportunities were in abundance. England ranked highly for employment and in opportunities for business. My focus was primarily on grabbing these opportunities with both hands. I was fortunate that when these opportunities arose, I was at the 'right place' at the 'right time' and my face fitted in. Instead of dwelling on the negativity and hardship that had engulfed my past; I decided to create a new life for myself, fuelled by energy and vigour.

In England, I immediately noticed that people hardly discussed religion. What's more, there were frontiers to promote equality in: 'gender, race and religion'. After fifty years of hard work and a satisfying career, I am now retired.

The last two decades have delivered turmoil and disaster across the world, prompting the question of where the world is heading. The nightmares that consumed my childhood began to re-emerge, as frequent and repetitive episodes. Religious differences are underpinning the violence that is once again rife. The western attitude towards Muslims is hardening because of backlash to the 9/11 and 7/7 terrorist events.

Many in western society view the Muslim States as fanatical, violent, immoral and greedy entities[7]. Conversely[130], Muslims have developed a negative view of the west. They appear to blame western policies for their own lack of prosperity. The Muslim world also believes that collaborations of western countries may have seeded fanatical groups responsible for terrorism. This can be rationalised via the unjust policies of the United States.

The truth is that there is an obvious greed to secure more power on one side; and the inability to cope with corruption and poverty on the other. Again, 'religion' is being used as a tool to create a deep attitudinal rift between the Christian West and Muslim East.

Retirement has allowed me to spend my time in leisure. I have started to study and examine the above 'taboo' issues fearlessly, with 'no holds barred'! After a lengthy and logical scrutiny and a scientific analytical study, I believe that the universe was designed purposefully, and it represents a specific intellectual rationale, currently beyond the human comprehension[84]. Hence, in a world full of prejudices, we need to evaluate our own self-consciousness. A soft voice in my subconscious mind is telling me that we must have a profound respect for the cultural diversity of each other. We humans ought to live as peaceful and harmonious beings. I believe this could be achieved if we permit ourselves to acknowledge that some of our beliefs about religions are wrong. For example, 'biblical scriptures' have justified slavery, sexism, racism and allowed 'concubinage[223]'. Kings and emperors in the past have used religion as a tool to mobilise people to fight and commit atrocities.

In today's world, one can draw comparisons between major religions of the past, and the current influence of the media. Manipulation of religion by the kings and authorities of the past and the current manipulation of thoughts and beliefs through the media by the states are directly related. The state is using 'media' as a tool to motivate and influence people to change their beliefs, attitudes and behaviours. In the past couple of decades, genocide has taken millions of lives in Iraq, Libya, Syria, Yemen, Rwanda and Burma. Yet, the media continues to downplay the genocide of millions, instead, choosing to report stories that gratify western viewers' perception as being idealistic, just and civilized. The truth contradicts these prejudices. The barbaric and deliberate extermination of millions of people will not be reported unless it is in the interest of the western governments. 90% of the media is controlled by six corporations in The United States of America. These corporations follow the government policies as instructed.

Peace can only be shared and secured if there is an absence of hostility and retribution[8]. Peace needs understanding, tolerance, cooperation and integration. This requires policies of equality, openness and justice to be embraced by world powers. All that is evident is 'hypocrisy' and 'greed' (Global Resources, 2018).

Winston Churchill once quoted, "If the human race wishes to have a prolonged period of material prosperity, they have to behave in a peaceful and helpful way. Unfortunately, we are living in a very unequal world, plagued with poverty." (*The Telegraph*, 2016)

Human history has been forged[9] by choice and muscle power. Our earth is split into boundaries, ideologies and philosophies. Abstract forces in history have been scarce. It must also be acknowledged that climate change and natural disasters such as earthquakes, volcanic eruption and major flooding have also dictated the path of history. We can classify these events as non-human. Excluding this, humanity is responsible for its own destruction. The chief motive being the gain of more power and the control of other people.

CHAPTER 1

BEGINNING OF THE UNIVERSE

I am curious about a few provocative and perplexing[195] topics, such as, 'Who created the universe?' and, if a good and loving God exists, then why is there so much suffering in the world? However, the question arises, 'Are we the descendants of Adam and Eve as the scriptures of the biblical books reveal, or is the scientific cosmological theory of creation accurate?' I am hoping that between religious, scientific and philosophical analyses, lies the answer that I seek.

Religion is primarily about ethics, morality, spirituality and faith. When we try to use or apply religion to science we create problems. We then try to resolve them through philosophy. Science has mainly been concerned with the physical world and has emphasis upon empirical[111] testing. Religion, on the other hand, emphasises social and emotional aspects of how we should live our lives.

Philosophy's main concern is the use of reason to investigate the validity of knowledge.

1.1 PHILOSOPHICAL VIEW

Although not initially formalised as a study, philosophy is as old as the existence of humans. It can be argued that philosophy was born the moment the human beings began to reflect on such questions such as: Who am I? Why am I here? Where am I going? What is the purpose of life?

The earliest texts on Hindu philosophy describe the universe as going through repeated cycles of creation and destruction. It was these ontological[11] questions concerning existence, and what happens when we die, that made Gautama Buddha go on a quest to seek enlightenment[82] through the state of Nirvana. From there on, several prophets and philosophers have tried to piece together this puzzle of existence, and their ideas of the afterlife.

The distinctive history of philosophy begins with the philosophers of Ancient Greece. The genesis[242] of science began with the philosophies of Plato and Aristotle. Between the two schools of thought, the idea of deductive reasoning emerged, and this is the cornerstone of scientific methods.

When I compare the conflict between religion and science, this makes me consider philosophical approaches towards the world, as studied by Aristotle, Plato and Socrates. Aristotle believed that the universe was always there. He believed that any changes that occurred were due to the natural disasters which kept returning civilization to another beginning. He believed that different 'gods' were the prime movers of the universe.

The Platonic approach demands that new knowledge comes from the spiritual world. It deals with the things that are not obvious, but are related to the 'Five Senses'. In simple words, if we must ask, 'Does God exist?' a Platonic answer will respond with a philosophical argument for the existence of God. Plato's theory of 'gods' was that some entity must have created the world, the sun, the moon, the stars and other things. Since most people believe in 'gods' it was argued that the majority was unlikely to be wrong. Intellectual arguments for the existence of 'one God' had not been developed at that time.

Aristotle and Socrates would respond to the same question, using science and mathematics, however, Socrates did not believe in 'Homer's gods' as Plato did. He believed in the possibility of a divine order and identifies that as 'good'. Socrates visited the holy land 'Jerusalem' and with Israelites, studied philosophy. Amongst them were the prophets of that era. He acquired various principles of teachings and knowledge. After he returned to Greece, he founded the system known as 'Unity of God'. (Gottlieb, 1997)

Plato in his book, *Apology* states, 'I cannot find anything in Socrates' statements which contradicts his belief in one God.' Socrates claims to have contact with a divine force, which is also the claim made by prehistoric people as being the wisest and most trusted. However, it was Socrates who laid the groundworks for the Western system of logic and philosophy; he was consequently sentenced to death in 399 BC.

Abdul-Baha was one of the most prominent philosophers and theologians from Iran. In his book *The Promulgation[12] of Universal Peace* (page 406), he confirmed that the Jews in the reign of Solomon were renowned in their degrees of art, letters, and their refinements. Their fame was widespread. Baha further states, *"Even the celebrated philosophers of Greece journeyed to Jerusalem in order to study with the Israelite sages[125]"* and many were the lessons of philosophy and wisdom they received. Among those philosophers was the famous Socrates. He visited the holy land and studied with the prophets of Israel, acquiring the principles of their philosophical teachings.

Immanuel Kant, a central figure in modern philosophy, was a German philosopher (Philosophy, 2010). He points out towards logical contradictions. The theory, he called 'Thesis'. He argued that if the universe had a beginning, why did it wait for an infinite amount of time before it started? If the universe had existed forever, why has it reached the present stage in an infinite amount of time? Thus, anti-thesis, in his view 'time' was the most important factor.

Kant's doctrine 'transcendental idealism' points out a distinction between what we can experience in the observable natural world and what we cannot see, like God and soul. His contribution to metaphysics and ethics had a profound impact on philosophy for centuries. His philosophy of religion included the existence of God, the immorality of the soul and the problem of evil. His philosophy defined the relationship between immorality and religion.

It created tension between the Christian faith and scientific knowledge. Kant's moral theory is based on his view of human beings as having a unique quality of rationality. It is this propensity and rationality that requires human beings to act in accordance with and for the sake of moral law and duty.

It is said that philosophy begins in wonder! As such, the ancient philosophers wondered what constituted the basic foundations of the world around them. The puzzle of existence and its metaphysical[15] dimensions had been the concern even before the Greek philosophers. According to biblical teachings, Abraham was the first prophet to preach about one God 4,000 years ago and from there on, there is a long line of prophets passing the same message well before Socrates.

Bertrand Russell (1872-1970) elucidated[16] a relationship between the three domains as follows: All defined knowledge belongs to science. All dogmas[17] that surpass definite knowledge belongs to theology. But between theology and science, there is a 'no man's land': which is known as philosophy (Russell, 1945).

1.2 RELIGIOUS ASPECT

Most people, in this age of science and technology, believe that their faith and religious beliefs do not conflict with science. Yet when it comes to the theory of evolution[118] and the creation of the universe, conflict exists and is evident. 'The Big Bang' has a significant bearing on religion and philosophy. If we take the 'The Big Bang Theory' and 'Darwin's Theory of Evolution', and study comparisons of these theses with the biblical books, a conflict between faith and science cannot be avoided.

The story of creation is described in the beginning of the Bible in 'Genesis[18]'. The Qur'an's verses, relating to the stories of creation, are spread across the book. The creation narrative in 'Genesis' is comparable with the Qur'anic version, though some discrepancies exist.

The old Bible texts were mainly written in Hebrew, and later translated into Greek. The New Testament was originally written in Greek. For this reason, I have decided to make more references to the Qur'an, due to my inability to understand Greek or Hebrew. However, some references to these texts have been quoted by the use of the English Bible.

The Qur'an insists that 'Allah' is the sole creator of all things, the universe and everything in it. The Qur'an contains verses explaining that God created heaven and earth in six days. According to Surat-Al-Araf, "Heavens and earth were created in six yawm [days]". The book of Genesis states that the creation of the world took place in seven days. The seventh heaven is inhabited by God and the angels. The sun, the moon and stars follow a regular path. Human creation was started with 'Adam' who was created with clay. God himself formed the materials of which Adam is made and then breathed the spirit into him (the soul). Later on, Eve was created from Adam. There are multiple occasions where the Qur'an states that life and the living creatures originated from water. Intriguingly, the Qur'an, 1,400 years old, provides an accurate description of human embryology[19], providing a step-by-step analysis of this biological phenomenon. Postnatal events are also discussed. Modern embryology development began in the 17th Century, with the invention of the microscope. Keith L Moore, Professor of Anatomy at the University of Toronto, confirms that Surah 23.13, 23.14, 22.5 in the Qur'an are scientifically correct.

However, many classical Muslim scholars, notably Al-Biruni, assigned[20] to the Qur'an a separate and an autonomous[21] realm[22]. Al-Biruni holds the view that the Qur'an does not interfere in the business of science. Many medieval scholars argued that there are multiple scientific explanations of natural phenomena[23] and refused to subordinate the Qur'an to an ever-changing science. In other words, whilst scientific theories change and develop, the word of the Qur'an should remain constant.

Sayyid Qutb, a Muslim scholar, states: "It does not take notice of the fact that scientific knowledge and scientific theories are always incomplete and are provisory[24] by the very nature. Therefore, the derivation[25] of scientific theories and knowledge in the Qur'anic verses are actually Testament to limiting the validity of these verses." (Qutb, 1964)

In biblical books, God has empowered humans to prevail as good over evil, compelled to become content and obey the commandments[107]. Stay away from pride, envy, wrath, greed, lust and malice. Feel compassion[96], empathy and sympathy. Lead your life to your true human potential. Abraham was the first person (prophet) to touch upon the idea of one God. In the Jewish Torah, the story of Abraham is narrated. It discusses human obedience to the will of God.

Jews were the first monotheistic[135] people. Christianity and Islam are the split-off groups from their Jewish roots. Abraham is known as the father of nations, because Jews, Christians and Muslims are all his descendants. After Abraham, there were several major and minor prophets. Prophets were interpreters of the messages of God, i.e. when God's spirit solemnly declares to men what he has received. The most prominent prophets include Moses, Isa (Jesus) and Mohammad. Half of the world's population follows the teachings of the above three.

1.3 SCIENTIFIC THEORIES

Most astronomers believe in the theory of The Big Bang; it is considered the prevailing[26] model for the beginning of the universe. It happened about fourteen billion years ago. At that time, the whole of the universe was inside a bubble, which was thousands of times smaller than a pinhead. Then it suddenly exploded with a Big Bang. Time, space and matter all began with this explosion. In a fraction of a second, the universe grew from smaller than a single atom to bigger than a galaxy and kept on growing at a rate which was faster than the speed of light. This was known as an inflationary era. As the universe expanded and cooled, energy changed into particles of matter and antimatter. These two opposing particles annihilated[27] each other. However, some matter survived. The most stable particles, known as, protons and neutrons, started to form when the universe was one second old.

Science has gradually accumulated evidence that the universe was created through a unique phenomenon. Science believes that The Big Bang was the creation of the universe. We do not know who or what caused The Big Bang, but scientists have calculated that the start of everything was instant, after the cosmic explosion, approximately fourteen billion years ago. The explosion from a singular atom started the universe, time, gravity, space, speed of light, scientific laws etc. It is logical to believe that God created everything at that moment. God can be viewed as a non-physical eternal and is present everywhere. Strangely, religious philosophical views on this beginning bear resemblance to their scientific counterpart. Divine prophetic statements are very similar, that the universe started with the light. The cosmic explosion of light could be compared with the biblical book's explanation, Genesis (1-2), in the beginning, God created the heaven and earth and said, "Let there be light, and there was light."

In 1915, Einstein introduced his 'Theory of Relativity' in which space and time were no longer absolute (Born, 2012). They were no longer a fixed background as the Greek philosophers had believed. He believed time and space were dynamic quantities that were shaped by the matter and energy in the universe. This raised philosophical questions about existence. The Theory of Relativity was proven with an atomic clock. Such a device was able to keep time to within one second in about 3.7 billion years. The clock illustrated a clear correlation[28] between time and altitude. In other words, time moved faster at a higher altitude.

In 1920, Edwin Hubble used a telescope at Mount Wilson to discover that stars were not universally distributed (Christianson, 1996). They were in vast clusters referred to as galaxies, and by the measurement of the speed of light, he was able to determine their velocities. He also found that all the galaxies are moving away from us, indicating that the universe is expanding. Subsequently, if the galaxies are moving away now, they must have been together in the past. The universe was changing with time, expanding as the distance between the galaxies increased as a function of time. These observations backed up The Big Bang Theory, thus physicists believed they had answered the question as to how the universe began through cosmological observation. Theologians, through their divine and prophetic statements, claimed that God was a metaphor[29] for the cosmos.

After three billion years, the universe had cooled to roughly 3,000 degrees Celsius. Protons and neutrons started to form hydrogen and helium nuclei[30]. The heat energy released from these fusions smashed atoms together; protons and neutrons collided to make isotopes[31] of hydrogen. This is known as the era of combination. During this era, atomic recombination resulted in the release of transparent electrically produced gas, which is detectable today as cosmic microwaves[32] (known as background radiation). Four hundred billion years after The Big Bang, the universe began to emerge from the cosmic dark ages. The gravitational pull and small fluctuations in the density of matter then gave rise to vast web-like structure of stars and the emptiness seen today. Dense regions pulled more and more matter through gravity, forming stars, galaxies and planets. The normal matter, which includes all the visible stars, galaxies and planets, makes up less than 5% of the total mass of the universe.

Dark matter accounts for roughly 27% of the universe; 68% of the universe is dark energy. Everything on Earth, and all matter that has been measured using instruments, can be viewed as 'normal matter' catering for 5% of the universe. One version of Einstein's gravity theory contains the cosmological constant, which refers to empty space possessing its own energy. This form of energy would cause the universe to expand faster. Black holes cannot be directly observed with telescopes that detect X-ray light or other forms of electromagnetic radiation. Black holes emit powerful gamma rays, having devastating effects on neighbouring stars (NASA Science, 2018).

In recent times, the prevailing model of the Big Bang Theory has been proved correct by Professor Higgs' discovery of the Higgs Boson Particle (ATLAS Collaboration, et al., 2012). This particle is a quantified manifestation[33] of a field that generates mass through its interaction with other particles. To prove this theory, a huge financial investment was made to build a Large Hadron Collider (LHC) in Switzerland. The LHC is designed to smash sub-atomic particles (e.g. protons) into each other at velocities close to the speed of light. Beams of protons are shot in opposing directions through a ring-shaped tunnel which is seventeen miles long. In 2013, scientists confirmed Higg's theory and the Higgs boson particle was isolated, known as the so-called 'God Particle'. Astrophysicist David Spergel has described cosmology as a historical science, because when we look out into space we look back in time due to the finite[69] nature of the speed of light (Spergel, 2015). Cosmology was shaped through mathematics and observations in an analysis of the universe. It is fair to say that cosmic inflammation was the direct result of The Big Bang.

CHAPTER 2

KNOWING THE CREATOR

So, the universe started with one enormous explosion of energy and light. That was the singular start of everything that may have existed. The Big Bang was the start of time, space and the universe. What eludes[34] this theory is, what caused that phenomenon to happen? Questions arise as to why the universe is so orderly and so reliable. Why do laws of physics remain constant? The earth rotates in the same 24 hours, day follows night. There was no logical necessity for a universe to follow rules. Why is 'nature' so mathematically correct? The fact that rules have always existed is a kind of miracle. Everything happens with intent, as if a creator has written a manual with purpose. Every cell in our body follows a detailed instruction code: the DNA code. Each of our cells contains a genetic code,[35] made from four bases. These are: adenine, thymine, guanine and cytosine. These bases are abbreviated to

A, T, G and C respectively[36]. There are three billion chemical bases in each cell. The code instructs the body to develop in a certain way. One could argue that you cannot find precise information like this without someone constructing it.

The physical cosmos shows distinctive patterns: matter is structured into galaxies, which are structured, in turn, into stars and the planetary system. On our planet, matter is structured in such a way as to produce living creatures. The question arises as to why the laws of physics and biology are the way they are. For example, gravity could have been a force of 'repulsion', not attraction. There would be no stable matter, and the development of life would have been impossible under these conditions. Does this not give us excellent reasons to believe in the existence of a creator? A tempting argument is that an intelligent being brought into existence 'the cosmos', together with its physical and biological laws. The only thing yet to be proven is the 'purpose'.

The earth remains a perfect distance away from the sun, while it rotates at a speed of 67,000mph. Around the sun, our moon is the perfect size and distance from the earth, dictated by the earth's gravitational pull. The moon creates ocean tides through a gravitational pull in such a way that ocean water does not stagnate, and yet, is prevented from spilling over across the earth. Taking water as an example, this colourless, odourless and tasteless liquid is a fundamental requirement for the survival of any organism[104]. Plants, animals and human beings consist mostly of water, it is a universal solvent, promoting the transfer of various chemicals, minerals and nutrients throughout our bodies and into the smallest blood vessels before their diffusion into the surrounding cells. Water has a unique surface tension; in plants, it can therefore flow upwards against gravity. At temperatures below 0 degrees Celsius, its unique solid crystal structure results in a density that yields a frozen surface beneath which liquid water persists. Fish, and other marine life, can thus, survive.

Another wonder of nature is the human brain, with its ability to manifest[143] emotions, thoughts and store memories. The capacity to simultaneously track involuntary bodily functions, such as breathing, hunger and digestion, make it a truly remarkable natural instrument, with our understanding of its mechanisms of action still at a primitive stage.

People who have faith don't require any explanation about the creator. Faith is deeply rooted in the hope of good things to come, especially when we must deal with life's monumental[37] failures. When hope starts to fade away, faith steps in. Whilst the concept of hope lives in the mind, faith is steeped in the heart and soul. It cannot be understood through a single dimension. As the oxygen in the air nourishes the body, faith nourishes the heart and soul. It is the energy that courses through every fibre and cell within our being. It is the fundamental foundation of our existence.

Some people do not believe in things they cannot see, but there is an enormous level of importance attributed to having faith in life. It is easy to allow stress, anxiety and fear to run our lives. We go from moment to moment worried, leading to highly stressful situations. This causes not only mental anguish but physical problems as well. There is a clear and documented connection between stress and the increased likelihood of disease and illness. One could hypothesise that we can learn to harbour faith and use it to eliminate stress, anxiety and fear.

At the same time, it is good to understand the cosmological argument that the universe has a transcendent[38] cause for its existence. The description of the creator is based on real properties of the universe. If God is described as omnipresent[39], omnipotent[40], transcendent, infinite[41], eternal[42], a creator and designer – then all these descriptions are true of the cosmos and religion. Both promote the creator from a single atomic explosion. Science provides much stronger evidence that God is the metaphor[29] for the cosmos[43]. God has no beginning or end in time or space.

One can assert that the creator of the universe has surrounded us with evidence of himself. Hazrat Ali said, "Man is a wonderful creature; He sees through layer of soft fat (Eyes) hear through a bone (Ears) and speaks through a lump of flesh (Tongue)" (Ali, 10th Century).

However, we must still find answers through science, with the aim to understand the development of life from non-life. A biogenesis[44] at some point happened, whether this be in space, in the ocean or in the atmosphere. The complexity of this process is evident in the fact that we have been unable to create life from non-life in the lab, although we have taken some amazing steps in recent decades. Science is yet unable to prove or disprove the existence of a creator, but if we use our beliefs as an excuse to draw conclusions that we are not scientifically ready for, we run the great risk of depriving ourselves of what we might have come to truly learn. I have tremendous respect for those who are believers in their faiths. Faith does not require that science gives a specific answer for their beliefs.

CHAPTER 3

EVOLUTION

3.1 RELIGIOUS ACCOUNT

According to the oldest religion of the world, Hinduism, 'Brahma' takes care of all evolution[118] and the creation process. According to Upanishads, there was just the soul. Then, it divided itself into 'male and female', 'the soulmates'. In a similar way, everything that exists is in pairs. If there is Shiva, there is Parvati. If there is light, then there must be darkness. This law of duality exists in almost everything. For every negative there is a positive. There are protons and there are electrons. As per the theory of evolution, humans have evolved via a gradual process, forming a chain of species, linking back to when all organisms[104] were purely aquatic, i.e. fish. This is stated in the

'Dashavatara'.

Hinduism preaches that all knowledge pre-existed to be recovered rather than discovered. The philosophical system of 'Samkhya' is noted to have both a religious and scientific context. Topics range from the theory of human evolution, to physics and biology. According to a survey conducted by Pew Forum in the United States, 80% of Hindus agree with the theories of Darwin's Theory of Evolution. Hindus had already believed in a notion of common ancestry between human and animals through 'Dashavatara'. It claims that our universe had many births and deaths.

In the biblical text, according to Genesis, Adam was immediately made in the image of God and Eve was made from Adam. Both were created fully formed. Mankind therefore occupies a unique position and has no direct link or relationship to any other creation. Theistic evolutionists believe that God used the process of evolution to create all living organisms[104]. It can be argued that God has planted evolution through a single-cell life that has undergone the process of evolutionary change and development.

Recently, theistic evolution has grown in popularity due to its acceptance by some leading evangelical pastors and theologians. They would assert that the six days of creation in Genesis were not normal days and in Genesis 6:8 NOAH'S flood was a local event rather than a global event. They also believe that there were human-like creatures that existed before Adam and Eve.

For some people, an evolutionary account of the origins of humanity may be a challenge to their religious faith. In principle all biblical religions are 'creationist[45]', whereby people believe that the sacred text provides an inerrant account of how all life and mankind come into existence in the universe; namely in 'six 24-hour days', some 7-10,000 years ago.

Al-Azami, Usaama 'Muslims and Evolution in the 21ˢᵗ Century – A Galileo Moment' (Huffington Post Religion Blog. Retrieved on 19ᵗʰ February, 2013)

'Old earth creationism' is the belief that the sacred text is an infallible[46] account of why the universe, all life and mankind came into existence, but accepts that days of creation are metaphorical[29] and could represent a very long period. Yet, they believe that all origins of life are the direct consequence of distinct acts of divine intervention.

'Progressive Creationism' is the religious belief that God created new forms of life, gradually, over a period. That period stretches over millions of years. It accepts mainstream geological and cosmological estimates for the age of the earth. Such progressive creationists believed that creation occurred in rapid bursts from which all kinds of plants and animals appeared in stages, over the course of millions of years. These bursts represent instances of God creating new types of organism[104]. They believe that species do not gradually appear by steady transformation from their ancestors.

(humanorigin.si.edu/about/broader-social-impacts-committee/science-religion-education-and-creationismprimer)

Muslims believe that the Qur'an is the literal word of God and it contains scientific information that would only have been discovered by the world in modern times, thus proving its divine origin centuries after its revelations. Scientific information pertaining to creation, astronomy, biology and human reproduction is spread in different 'Suras'.

The first mention of the creation in the Qur'an is in Surat-Al-Anbiya. It states that the universe was joined together as one unit, then Allah demanded that planets and stars were formed. He who created the sun and the moon (celestial bodies) swim along the skies, each in its own orbit. In Surat-Al-A'raf, it is explained that the heavens and the earth were created in six 'yawm'. Each 'yawm' is equivalent to 50,000 years. The characters of Adam and Hawwah (Eve) appear in the Qur'an as the first man and woman; however, how these two people developed is not explicitly described.

The Qur'an refers to Adam and Hawwah as the first man and woman. The Qur'an explains that Allah created Adam with his own hands, and consequently Eve was created from Adam. Islamic scholars proposed that the verses could have various interpretations. Therefore, it is open for speculation as to whether Adam and Eve evolved naturally from a common ancestor or whether they developed because of supernatural[77] forces, created by the miracle of Allah. This is discussed in *Prophets in the Quran* by Brannon M. Wheeler.

Islam also has its own school of creationism, known as 'theistic evolutionism', which claims that mainstream scientific analysis of the origin of the universe is supported by the Qur'an. Dr Khalid Anees, of the Islamic Society of Britain, states, "Muslims interpret the world through both Qur'an and what is tangible and seen. There is no contradiction between what is revealed in the Qur'an and natural selection and the survival of the fittest."

It is important to analyse the truth regarding evolution[118]. The scientific theory of evolution has led to the widespread acceptance of an atheistic, materialistic world-view. Faith teaches us that a spiritual world-view with God as being the intelligent designer is not scientific. In my opinion, the claim that Darwinian evolution has proven science over religion needs to be examined. This is because these theories provide no explanation as to how life originated in the first place. Darwin's theory is based on the twin principles of variation and natural selection.

Douglas Futuyama's book *Evolutionary Biology* published in 1979 from pages 254-271 states: 'it was Darwin's theory of evolution together with Marx's theory of history and Freud's theory of human nature that provided a crucial plank to the platform of mechanism and materialism that has since been a stage of most western thought.'

Christian fundamentalists dispute the evidence of a common descent. They argue for the Abrahamic accounts of creation. The Catholic Church now recognises the existence of evolution. Scientific study has provided us with a deeper understanding of both our connections to life on Earth and the uniqueness of our species, *Homo sapiens*[47]. Science has proven this through DNA testing.

The arguments of many theologians against scientific theories is that if these theories change under the light of new discoveries, why should we believe what it may change into tomorrow? Although this has occurred occasionally in the history of science. The media coverage of advances in science often sensationalise the nature of the discoveries and are also focused on the most controversial interpretation of new findings.

3.2 SCIENTIFIC THEORY OF EVOLUTION

In 1859 Charles Darwin published his *Origin of the Species*. It is one of the most controversial books of the past millennium. Evolution does not seem to turn many Christians into unbelievers. Darwin's theory explains the origins of the biological processes of evolution.

Simple (organic molecules), consisting of carbon, hydrogen, nitrogen, oxygen and phosphorus atoms are the building blocks of life on Earth. Experiments suggests that organic molecules could have been synthesised in the atmosphere of early Earth and then rained down into the ocean. RNA and DNA molecules were the genetic material for life, captured as long chains of organic molecules. All living things reproduce their genetic material and pass it on to their offspring. Therefore, the ability to copy a molecule that encodes genetic information is a key step in the origin of life, without which, life could not have existed. This ability was first evolved in the form of RNA, formed through the chain of nucleotides (organic molecules). Life relied on RNA for nucleotides. A bottleneck in evolutionary history is observed in the development of DNA, which contains instruction for RNA, becoming the main and fundamental characteristic of self-replicating material, which is present in nearly all living organisms[104].

DNA (deoxyribonucleic acid) is known as the tree of life. It has demoted RNA to the role of a messenger, which now carries information from DNA to protein.

The breakthrough tentatively solved the problem of the 'chicken and egg' question. Life switched from RNA- to DNA-based inheritance.

		HUMAN TIMELINE
1-	*Homo sapiens*[47]	Modern human – Modern speech
2-	*Neanderthal*[48]	Earliest clothes – Earliest cooking

		Earliest in Europe
4-	*Homo erectus*[49]	Earliest in Europe – Earliest discovery and use of fire
5-	*Homo habilis*	Earliest exit from Africa
6-	HUMAN-LIKE APES	
7-	HOMINOIDEA APES	
8-	HOMININE GIBBONS	
10-	GORILLA	

http://en.m.wikipedia.org.>wiki>history of human migration – Wikipedia and the history of human migration (book) by Russell King (ISBN: 9781845377960).

3.3 THE FIRST HUMAN MIGRATION

The Omo River in Ethiopia yields the earliest fossil[50] evidence for modern *Homo sapiens*[47]. The dominant model of geographic origin and early migration of *Homo sapiens*[47] from Africa took place in two waves. The first took place about 130,000 years ago, but they retreated back to Africa as their settlements were replaced by Neanderthals[48]. There is some evidence of a presence of modern humans in China 80,000 years ago.

The second wave of migration took place roughly 100,000-50,000 years ago, through a southern route. This led to lasting colonisation of Asia, Australia and Europe. These waves of immigrants interbred[51] with local *Homo erectus*[49] populations in multiple regions. Fossils[50] of early *Homo sapiens*[47] were found in the Qafzeh cave in Israel. These fossils have been dated to 100,000-80,000 years ago. Fossils from Lake Mango have been dated to 42,000 years ago.

Biologists have proven that there is only one human kind and it is recognised as *Homo Sapiens*[47]. These species have subgroups across mankind in skin cashing[52], ranging from the black skin of some Africans and Melanesian islanders to the very pale skin of Northern Europe. Every person's skin colour is a combination of the same pigments: melanin carotenoids and haemoglobin. Melanin is a dark brown pigment and haemoglobin is a red pigment in blood. Melanin is a specialist pigment produced in cells named melanocytes. All human races have a high concentration of melanocytes, but different skin tones result from variation in how much melanin these cells produce, i.e., skin pigmentation is dependent on how active the melanocyte cells are. In very dark skin, it is active all the time. Melanocytes are moderately active in mid-brown skin tones. This activity can be increased or decreased depending on numerous factors. Such factors include exposure to sun and changes in hormones.

After migration of the *Homo sapiens*[47], people of different regions developed different features; these included varying heights, limb proportion, head shapes etc. Such variation was a result of genetic mutations. Characteristics such as growth of bones, cartilage and muscle were adapted to support their local environment via the medium of genetic mutation. This divided the Mongoloid and Caucasoid[53] populations in the Asian regions (Indochina). Mongoloids decided to head east, circumventing[54] the Himalayas to the north. The Caucasoid[53] tribes split up. Some drifted southwards, to Iran and Pakistan, and became known as Indo-Aryans. Others migrated west, becoming modern-day Europeans.

The populations that were closest to the equator[55] eventually developed a darker skin tone than those who migrated and headed to Europe. The darkness of people corresponds to longitude, as does hair texture. Indigenous[56] populations that stretch along the equator share similar traits[209]. Distance from prime meridian[57] seems to correlate with variations in indigenous population. More easterly Asians have facial features like those of other populations with an almost equal distance from the meridian.

3.4 THE FIRST HUMAN COMMUNITY

Many different elements must come together before a human community with the required level of sophistication develops. This is commonly referred to as civilization. The roots of civilization reach back to the earliest introduction of primitive technology in the early Stone Age Palaeolithic Era: this was followed by the new Stone Age and Neolithic era. As humans began the systematic husbandry of plants and animals, agriculture advanced and most humans transitioned from nomadic to a settled lifestyle as farmers. The nomadic tradition of hunting continued in some locations, especially in isolated regions.

*(***Understanding the Neolithic*** by **Julian Thomas**, ISBN: 9780415207669, page 16)*

As farming developed, human communities expanded into larger units (between 8000-5000 BC). Settlements grew into cities on the banks of the rivers and lakes.

The first prominent and well-developed settlement arose in Mesopotamia, present-day Iraq). It was the land between the Rivers Euphrates and Tigris. The civilizations progressed from there onto the banks of the River Nile in Egypt and along the Indus valley as well as along the major rivers of China.

The rapid expansion of humans took them to Oceania[58] and North America.

(http://en.m.wikipedia.org.>wiki>history of human migration – Wikipedia, and **The History of Human Migration** *by* **Russell King, ISBN: 9781845377960)**

Humans gradually had colonized nearly all the 'ice-free' parts of the globe. Other hominids[119] such as *Homo erectus*[49] used wood and stones for hunting. At some point the humans started to use fire for heat and cooking (Harari, 2014).

Rivers had two advantages for the developing communities: they provided water for irrigation as well as being the easiest method of transport for societies yet to invent paved roads.

In the 'Early Stone Age', humans developed language and the systematic burial of the dead. Art also started to develop in the form of cave paintings and sculpture. Grain agriculture became more sophisticated and promoted a division of labour. Growing complexity in human society necessitated a system of writing and accounting. Continuous farming led to depletion of soil, resulting in the requirement of fertile land further afield. In river valleys, populations became dense due to annual flooding, which renewed soil fertility. People began to use animals, domesticating them for pulling ploughs and carts. They also used these animals to produce milk, meat and manure. The skin and wool of the animals were used for protection from cold.

As the societies grew, so did the killings, which were driven by personal motives, such as family disputes generally resulting from greed, jealousy and adultery. Groups became more territorial and communities living in unwalled villages become vulnerable.

The walled cities became associated with civilization and the beginnings of the accumulation of transferrable surpluses such as grain and cattle. It was the start of the state, whose function was to provide security against war and violence. A soldier class emerged through hunters, who provided security; in return, they were rewarded with land, grain and women.

In ancient Mesopotamia, there were three main classes of people. These included the upper class – powerful, wealthy and consisted of kings, nobles, priests and warriors; the middle class, made of people who were paid to work as farmers, fishermen, merchants and artisans[60]; and lastly, the lower class, slaves who performed all manual and domestic work. Slaves worked for the kings, priests and the wealthy.

3.5 MORALITY AND GROUP LIVING

Franz De Waal and Barbra King both view human morality as having grown out of primate[61] society (De Waal, 2016). Therefore, the following characteristics are shared by humans: attachment, bonding, cooperation, mutual aid, sympathy, empathy, peace keeping and caring. These entities are what others may think about you. The above became a means of social control and group solidarity[59]. Society's moral codes were enforced rigorously with reward and punishment.

Since the ancient civilizations, religion and morality have been closely intertwined.

Historically, morality and religion have been inseparable, that is, until very recent times. The term 'moral' comes from the Latin 'mos' which means habit or custom. It is translated in Greek as 'ethos'. In English, there is a non-technical use of ethos as ethics. For example: medical ethics.

Psychologist Matt J Rossano agrees that religion emerged after morality by expanding social scrutiny of individual behaviour to include a supernatural agent (Rossano, 2010). Religion requires a system of symbols for communication, such as language, to be transmitted from one individual to the other. Under these conditions, religion cannot pre-date the emergence of language. Religious ideas and oral traditions were articulated by shamans[62], where a written text was maintained by a select group known as 'the clergy'. Shamans are defined as the people who could communicate with the spirits and could heal illness caused by evil spirits. By including ever-watchful ancestors, spirits and gods in the social realm[63], humans discovered an effective strategy for restraining selfishness. The adaptive value[64] of religion would have enhanced group survival.

Social order is maintained by certain rules. Since community size grew over the course of the human evolution, greater enforcement of rules was required to achieve group cohesion. Morality may have evolved as a means of social control, conflict resolution and group solidarity[59]. Society's moral codes were enforced rigorously with reward and punishment.

The shift from hunter-gatherers to agricultural subsistence gave rise to social stratification[65] and population growth. As humans started to form communities around 'agricultural' land, abstract power structures started to arise from the division of labour. As the number and size of the agricultural societies increased, they expanded into land used by hunter-gatherers.

As population density rose, the struggles for surplus resources evoked the worst in people. Violence and conflict, intimidation and rape, plunder and conquest offered more rewards than the quiet cultivation. The winners would become chiefs, kings and emperors. They monopolized the biggest houses, treasures, precious stones and most beautiful women. They wrote laws to reinforce a hierarchy[122] based on status, such that it suited them. That emergence of horticultural and pastoral[66] societies led to social inequality. Horticultural societies cultivated plants, while pastoral societies domesticated and bred animals.

Division of labour in agricultural societies led to job specialisation and stratification[65]. People began to value certain jobs more highly than others. Manual labour became the least respected, while those engaged in high positions became most respected. People began to trade goods and services and began to accumulate possessions. Some accumulated more than others and gained prestige in society. Possessions were passed to the future generations, concentrating wealth into the hands of a few groups. The unlucky losers became slaves. Systematic inequality was reinforced by chiefs and kings. Townships grew into cities like Jericho and Huyuk. The leaders of these cities were throwing themselves into warfare against their neighbours. These early states became criminal enterprises having found the ideal way of perpetuating[67] and legitimising themselves. The state manipulated morals such that good and evil became defined by what suited their corresponding leaders.

To avoid constant warfare, those predatory[68] rulers sometimes came to an agreement about boundaries. They had the power to dominate people within those boundaries. Misbehaviour against the state was punished. New words were used to describe kings like 'Pharaoh' in Egypt, 'Shang' in China. They were recognised as the ultimate authority. They imposed binding rules. Those rules were run by special group of officials to collect taxes. This group had mastered the art of numeracy, and writing began via the system of the pictograph. Text was written on clay tablets. Sometimes the rules were narrated by symbols which represented ideas.

A variety of positions are apparent regarding the relationship between religion and morality. Some believe that guidance from religion is a necessity to a moral life. Others believe that it is impossible to distinguish evil from good unless one has a finite[69] reference point. This reference point is known as religion. Traditionally, the most important link between religion and ethics is to provide a reason for doing what is right. The reason is based on reward and punishment thus, those who are virtuous, will be rewarded with an eternity[70] of bliss, whilst the rest will be sent to hell.

However, one can argue that a man's ethical behaviour should be based on social ties, sympathy and caring. No religious fear of punishment or reward after death should become the base of morality. It is the motives of the individual that is of most significant.

In principle, the concepts of morality and religion are two different kinds of value system. There is nothing to stop atheists believing in morality, or human goodness. Many atheists lead more moral lives than religious believers, who confuse divine law and punishment with right and wrong.

CHAPTER 4

GOOD AND EVIL

When war, disaster or crime takes innocent loved ones, destroys their homes or brings untold suffering to them, people want to know why this suffering and tragedy befalls them. Why does God allow such suffering? Normally, the answer is that 'it is God's will'. The Bible answers it in revelation 12:9, "Satan is hateful, deceptive and cruel". Accordingly, the world under his influence is full of hatred, deceit and cruelty.

Humans can make choices. What is good or bad may be debatable. Each society has set codes and rules which define those perimeters. Faith in God gives comfort and hope. A major problem visible in religions, is observed in the hypocrisy with which they preach 'loving thy neighbour', whilst clandestinely[71] harbouring resentment and hatred against other religions. A religion may preach positivity; however, thousands of denominations

42

have developed over centuries differing from one another on trivial issues. This illustrates the disharmony that has resulted from making statements which are not based on scientific facts; corruption within religious institutions and the enactment[72] of senseless laws with the purpose of acquiring and sustaining power.

Even in the oldest of the religions, good and evil exists. If we believe that God is almighty and perfect than why does this evil exist? It seems as if God wants to obliterate[73] evil and is not able to do so, thus his almightiness is questioned. If God is not able to eradicate[74] evil, then his goodness should be questioned. The problem of evil mostly applies to monotheistic religions, such as Judaism, Christianity and Islam. These religions believe in a singular God that is omnipotent[40], omniscient[75] and omnibenevolent[76]. The problem of evil has been extended to non-human lifeforms, to include animal suffering from natural evils and human cruelty against them. This can be illustrated by the idea of a lion hunting down a deer to quench his hunger. To the human being, this be cruel, yet the lion is unaware of any sense of evil from its actions. Therefore, perhaps nature permits these outliers within the environment.

Hinduism goes back to 5000 BC, closely related to Jainism and Buddhism. According to Hindu texts, Lord Brahma is the creator of the universe. The other two gods are Vishnu, who sustains the creation, and Shiva, the destroyer of evil; these three Gods combine to create or destroy the universe. The hymns of Rigveda, the oldest scripture of Hinduism, mention many deities[124]. However, Ishvara is viewed as a special being to anything that has a 'spiritual significance'.

(www.ancient-origins.net>hind-sacred...text-about-human-origin-oo66).

Across the history of humanity, sanctuaries and places of worship were created to perform rituals and prayers. In 'Abrahamic' religions, the common theme is that good should prevail and evil should be defeated. In cultures influenced by Buddhism, both good and evil are experienced and perceived as part of an antagonistic duality that itself must be overcome through achieving 'Sunyate', that is, recognition of good to overcome evil. Buddha thought 'awakening' comes through one's own personal experiences and not through teaching.

In 'Abrahamic' religions, evil is a supernatural[77] force, which should be overcome by self-control and worship. Regarding the origin of 'good', it expresses the right and desirable qualities as decided by authority and society in general.

In the currently recognised version, the seven sins are: pride, greed, lust, envy, gluttony, and sloth. Each of these sins are a manifestation[143] of our ego (of self).

According to the Catholic Church, sins are divided into two categories – venial sins (minor) and mortal sins (deadly). The Catholic Church also recognises seven virtues to overcome the seven deadly sins.

Sins	Remedy
Lust	Chastity
Gluttony	Temperance (Self Control)
Greed	Generosity (Charity)
Sloth (Laziness)	Diligence
Wrath	Patience
Envy	Kindness
Pride	Humility[93]

The seven deadly sins have been a source of inspiration for writers and prophets: Aristotle, Confucius, Buddha, Dante, Voltaire, Iqbal and so on...

In Islam, the Qur'an's concept of the ego occurs many times. 'NAFS' is the agency of 'free will' and intelligence, without which neither responsibility nor accountability can exist. According to the Sufi philosophies, the 'Nafs' and its unrefined state in 'The Ego' is the lowest dimension of a person's inward existence. Nafs is inspired by your heart, sees the results of your actions, agrees with your brain, sees your weakness and aspires[78] to perfection. The Qur'an does not attribute any inherent properties of good or evil to the Nafs or self, but instead conveys the idea that it has to be nurtured and self-regulated, so it can progress into becoming good, through its thought and actions.

The Qur'anic concept of the Nafs therefore has an extremely modernistic undertone, much like Nietzsche's conception of 'Ubermensch' or

'Superman', as suggested by Muhammad Iqbal, a prominent Muslim scholar and philosopher. Iqbal stated that it is probable that Nietzsche borrowed his idea of Ubermensch from the literature of Islam.

Islam, like the Roman Catholic Church, classifies sins into two categories. The two types of sins are: Kabira (greater sins) and Sagheera (pardonable) sins. The Qur'an instructs its followers to control Nafs (inner desires) by the following virtues:

Vice	Translations	Remedies	Translation
Takabur	Pride	Salah	Humility[93]
Hirz	Greed	Zakat	Charity
Hasad	Envy	Detachment from material world	
Shahwa	Lust	Fasting	Self-Restraint
Bokhal	Stinginess	Frugality[79]	
Gheeba	Backbiting	Jihad	Diligence
Keena	Malice	Kindness & Love	

In Hindu theology, there are six passions, which prevent man attaining salvation. The deadly sins are those transgressions[116] which are fatal to spiritual progress.

Vice	Translations	Remedies	Translations
Kama	Lust	Sattvam	Purity
Krodha	Anger	Sama	Self-Control
Lobh	Greed	Dhama	Discipline
Moha	Temptation	Vairagan	Detachment
Mada	Pride	Sattyam	Truth
Matsarya	Envy	Ahimsa	Non-Violence

In Buddhism there is no concept of sin. Buddha thought that the 'awakening' comes through one's own personal experiences and not through teachings. However, the following offences and traits[209] deliver the offenders to 'Naraka', hell.

The five rules (precepts) constitute the basic code of ethics, undertaken by the followers of Buddha. These five precepts are the commitments to abstain from:

1. *Harming living beings*

2. *Stealing*

3. *Lying*

4. *Sexual misconduct*

5. *Intoxication*

Any follower of Buddha must give his commitment to the following four rules:

1. *I undertake the training rule to abstain from killing.*

2. *I undertake the training rule to abstain from taking what is not given.*

3. *I undertake the training rule to abstain from sexual misconduct.*

4. *I undertake to abstain from taking fermented drinks that causes heedlessness.*

In Judaism, humility[93] is among the greatest of virtues and is the antithesis[80] of pride.

	VICES IN JUDAISM
Be Just and Truthful	Unjust activity
Be Honest	Falsehood – fraudulent dealings
Don't Lie	Perjury
Be Kind	Don't be an oppressor
Be Charitable	Stealing
Eat Halal	Don't eat dangerous food
Give 10% of your income away	Stinginess
Speak Well	Evil tongue
Be Benevolent	Unethical sexual activity

Halakha is the Jewish law from the written and oral Torah. It is a law to direct a Jew to behave in every aspect of life, encompassing civil, criminal and religious laws. God gave Moses 'Ten Commandments' in the Torah.

The Ten Commandments:

1. *You shall not have another God, before me.*

2. *You shall not make idols.*

3. *You shall not take the name of the lord in vain.*

4. *Remember the day of Sabbath. Keep it holy.*

5. *Honour your father and mother.*

6. *You shall not murder.*

7. *You shall not commit adultery.*

8. *You shall not steal.*

9. *You shall not covet.*

10. *You shall not bear false witness against your neighbour.*

CHAPTER 5

FAITH AND SPIRITUALITY

There are many gifts that science brings to faith and spirituality, even though they are often seen by many as a constant conflict. The harmonious coexistence of science and faith could, in principle, create a more peaceful and habitable Earth. Science unearths facts that tell us more about the universe and its creator. For instance, science, via experimentation, unveils such ethical challenges of our time including climate change, pollution limits and so on. Faith and spirituality ought to be about righteous actions. Spirituality can be source of positive energy. As Martin Luther King Jr said, "Our scientific powers have overturned our spiritual powers. We have guided missiles and misguided men." For a real believer of faith, there is no need to prove in any rational way that God exists. The universe and the nature is self-evident in its creation.

Faith can be defined as believing in something without evidence of proof. Faith, on the other hand is a supernatural[77] term. From infancy every human has an innate[81] sense that there is something more than 'me'. The baby calls out for the mother, even when the mother is gone. There is a domain of thought on which human existence is governed. This domain of thought has been historically developed in connection with morality and faith. The faith involves a belief that makes an implicit reference to a transcendent source. Most faiths are based on interpreting the sacred pronouncements, either through oral traditions or in canonical writings, backed by a divine authority.

Some scholars claim that it is a mistake to believe that faith requires evidence. Religion has survived, because it helped us to form increasingly larger groups which were held together by a common belief. Ancient beliefs show that our basic cognitive[83] comprehension[84] biases us towards a thinking based on pre-life and after-life. There are people who think that faith and reason are unrelated. Thus, faith becomes a non-rational[85] belief. The doctrine[86] that knowledge depends on faith is known as Fideism[87]. It is an exclusive belief in faith alone and is independent of reason. As such, reason and faith contradict each other. Johann Hamann promoted the view that elevated[88] faith is the only guide to human conduct. He argues that everything people do is ultimately based on faith. In religious beliefs, faith may equate to confidence based on intuition[89].

One should face the paradox[91] that relates to the

transcendence[92] of life, accept the mystery of life and consequently readopt a faith system. Based upon this, I believe that only then can one enter spirituality or enlightenment[82], whereby you come out of the existing systems of faith and accept the universal principles of compassion[96], love and humility[93]. It is at this stage of faith that a Muslim put his trust in the complete submission to the will of Allah, which is known as 'Iman'. It must be accompanied with righteous deeds. Spirituality comes through the remembrance of the creator with gratitude. One could argue that one's faith may be based on the experiences one encounters with the community and the society one grew up in. In other words, a person's faith is often determined by their community. For example, if one is born in a Muslim family, one's parents are unlikely to take the person to church to listen to a priest's sermon[94].

Spirituality makes a person focus on one's faith. For example, through places of worship and numerous actions communities do to form a connection with God. In Muslim communities, daily prayers play a significant role and they believe that the purpose of existence is to worship Allah. Christians believe that only the believers of their faith will go to heaven and non-believers will go to hell. Fundamentalists in all religions believe in the doctrine that the God's laws and its source of declaration to mankind, is true and should not be questioned. Conversely[130], historians and even biblical evidence shows that to not be the case.

Spirituality can have different definitions and there are many ways to express one's spirituality.

Rituals, songs and dances are all methods of expression. Being spiritual could make someone feel enlightened[82] and could fill someone with strong emotions and deep feeling. In some cases, these emotions could be so strong that people would be willing to sacrifice anything to help maintain the integrity of what they find to be spiritual. There are countless people willing to sacrifice money, material goods, jobs and several other things that many keep dear to themselves, only to help preserve what they find spiritual.

Most researchers don't believe that the cognitive[83] tendencies that are biased towards religion, evolved specifically thinking about the afterlife. This tendency set us up to believe in an omnipresent God-like concept. Religion has evolved through morality as a means of binding people into large moral communities. Through stories and rituals, religions have built on five basic moral foundations: do not harm, play fairly, be loyal to your group, respect authority and live purely. Rituals helped people and allowed them to live cooperatively. The problem is, the more you look inwards towards your religious group and its claims of virtue, the less you look outwards, and the more distrustful you are of others.

This distrust has caused much of the world's strife and violence. It is true that religion has been a major feature of some historical conflicts and the most recent wave of terrorism. Religion has taken on extra significance because of ill-directed globalisation, greed and arrogance of superpowers. Sometimes I wonder who is more corrupt: the state themselves or fundamentalists?

5.1 AFTERLIFE

The scientific view of death is the cessation[103] of all biological functions that sustain a living organism[104]. Religion is a system of beliefs and practices that bind people with divinity[105], the ethics of 'good and bad' as a common dichotomy[106], related to 'reward and punishment' in the afterlife. However, physics and biology avoid supernatural[77] explanations to describe reality. They continue to consider science and spirituality to be complementary[100] rather than contradictory.

In Hinduism, a concept of heaven and hell exists. They believe in an afterlife and rebirth. They believe that at the time of death, the deity of death, either Yama or his assistants, visit and take away the departed spirit, known as 'Atma'. The account-keeping deity, Chitra Gupta, checks the deeds and misdeeds of the Atma. Then heaven or hell is assigned. The universal concept of 'Karma' still applies to whatever name you want to give it. It simply means, *'you will reap what you sow'*.

Many cultures and religions have the idea of an 'afterlife' and relate that idea to reward and punishment for past sins. In Buddhism, being reborn as a human being is considered a state in which one can attain enlightenment[82]. Therefore, death helps remind that one should not take life for granted.

Afterlife is a reward or punishment for conduct during life.

Coming back to biblical religions, an idea explained in many visionary documents is called 'Apocalypse[99]'. None of these were included in the Hebrew Bible, except the book of Daniel. Many scholars agree that the 'Day of Judgment' and the afterlife exists. Many Jewish writers have been influenced by their contact with 'Zoroastrianism', which saw the existence of an eternal struggle between the great principles of 'good and evil'. In their view, the dead set off across a great bridge towards the afterlife. For those who lived well, this was an easy crossing; the wicked ones were dragged down to an abyss[101] of darkness.

The idea of 'Heaven and Hell' was borrowed from another religion, it is only preached in the final 30% of the Bible. The concept of hell was not part of the Jewish traditions. It evolves into the Bible beginning with the time of Daniel. At that time, Jews were living in captivity of the Persians, whose faith was Zoroastrianism. Judaism is believed to be the first religion that incorporated[95] a concept of heaven and hell. Grolier's *Multimedia Encyclopaedia* quotes, "some elements of Persian religions were incorporated[95] into Judaism, a more elaborate[98] doctrine[86] of Angels, the figure of Satan, and a system of beliefs concerning the end of life and end of time". Whatever the argument of heaven and hell is, it is a part of superstition, detaching human beings from real practical spirituality.

The Bible originally did not have such a model.

Most religions have a concept of heaven and hell, although the names and symbols may be different. On a deeper level of spirituality, heaven and hell disappear as concepts. Through morality, compassion[96] and spirituality, we try to achieve higher rewards.

In Judaism and in Christianity, traditional heaven and hell came from the books of the Old Testament, verified by the Qur'an. In Abrahamic traditions, the dead tread a specific plane of existence after death; generally seen as an existence of supernatural[77] happiness or an empty abyss[101] of darkness, until the day of judgment arrives. Muslims call this existence 'Barzakh[102]'. It can be described as a kind of cold sleep, where the soul rests as it awaits the Day of Judgment.

In Judaism, death puts an end to the possibility of fulfilling any commandments[107]. Among Christians the death and resurrection[108] of Jesus are two core events. They believe that the lord's death on cross had paid the penalty for mankind's sins. They also believe in resurrection and 'The Day of Judgement.' Death in Islam is the termination of worldly life and the beginning of the afterlife. According to Islamic beliefs souls may rest in Barzakh[102] until final resurrection on the Day of Judgement.

Al-Ghazali in his book: *Faysal Al-Tafriqa-Baynal-Islam*, states: "The rule of reconciling conflict between reason and outward meaning of the revelation in that human life is a path towards judgment day and the reward and punishment gained from it". In another book, *Remembrance of Death and the Afterlife* (Ihya' ulum al-Din), Al-Ghazali gives a detailed account of how to remember death and teaches us how to use our time wisely whilst living. It focuses on subjects of intention, sincerity and truthfulness (Jackson, 2002).

There is one universal truth: we all must die, we all must make the journey and we all must cross over. The concept of the afterlife is comforting to those that are involved in acts that their society considered rewarding and divine.

'Afterlife' is a controversial subject and has been treated with <u>scepticism or cynicism</u>[109]. Science is still investigating and conducting research into the subject of 'after life'. Research among cardiac arrest patients has been conducted by Southampton University. (www.telegraph.co.uk on 7th October 2014). *The Telegraph* reported 'Some cardiac arrest patients recalled seeing a bright light, a golden flash or the sun shining.' The newspaper talks about the first hint of life after death in the biggest ever scientific studies. 'In that study they revealed near death and out of body experiences. They discovered that some awareness may continue even after the brain has shut down completely.'

Over the last few decades there has been a significant increase in research into quantum physics[110]. Some of the physicists working in this area are discovering no conflict between physics and paranormal events. After all, we know now that atoms are 99.9999% empty space. According to quantum physics the sub-atomic particles (electron, proton and neutrons) are not hard mass. They are made up of energy, thus the world we live in is not solid; it is in fact almost entirely empty space.

Dr Robert Crookall in his book *The Interpretation of Cosmic and Mystical Experiences*: "undertook a systematic study and noted the consistency of evidences given by people who had out of body experiences, near death experiences and the communications of high level mediums obtain the reliable results." (Crookall, 1969.)

The advantage of anything scientific or empirical[111] is that given any formula or principle, keeping variables[112] constant obtains the same results. However, wherever there is an inconsistency between science and beliefs, inevitably science prevails, even if the beliefs have been around for thousands of years. Victor Zammit and his wife Wendy formed a foundation to collect scientific evidence for the survival of consciousness beyond physical death. Victor was a successful barrister and Wendy a psychologist. Together they gathered evidence though the revelations of Electronic Voice Phenomena (EVP) and Instrumental Trans-Communications (ITC) over thirty years. Evidences was also collected through hypnosis[113], holotropic[114] states and near-death experiences (www.victorzammit.com). Victor's foundation is a non-profitable organisation, with the purpose being to promote the realisation of an afterlife (Zanmmit, 2000).

The following are the findings of Victor Zammit:

1. *All humans survive physical death, irrespective of their beliefs.*

2. *At the point of death, we take our mind, with all its experience, our character, our ethics and 'spirit'.*

3. *There are different levels of 'spheres[115]' in the afterlife.*

4. *Higher spheres are too beautiful to imagine.*

5. *Hells for eternity do not exist. It was invented by man.*

6. *You can move from lower-level spheres to the higher-level spheres. Those who were most evil during their life are naturally attracted to the dark and lower sphere.*

7. *Once death occurs, there will be a feeling of enormous lightness.*

8. *Nobody judges you. You judge yourself and measure the effect your words and deeds have on others.*

9. *Atheist and agonistic non-believers can all attain higher afterlife spheres.*

10. *One can still learn spiritual lessons and help others less informed. Gradually they will transpire to the next spheres.*

11. *In the afterlife, there is no need to eat, drink or sleep. Communication is done through*

telepathy. The body can travel at the speed of light.

12. *Selfishness is the greatest transgression[116] against spirituality, together with cruelly and harassment[117].*

13. *Unconditional love is the most powerful force in the afterlife.*

CHAPTER 6

ORIGIN OF RELIGION

During human evolution[118] the hominid[119] brain tripled in size, peaking some 50,000 years ago. The religious mind is one result of a brain large enough to formulate religious and philosophical ideas. The neocortex[120] of the brain is associated with language, self-consciousness and emotion. Some scholars have suggested that religion is genetically 'hard wired' into the human condition, known as the 'God gene'. In addition, dopamine[121] plays a major role in reward-motivated behaviour. Dopaminergic function promotes an emphasis on distance, space and time, which in turn relates to God, who is all rewarding.

Religious sentiment was played upon by shamans on behalf of monarchical advantages to control people's behaviour. In ancient Sumer, the city was not built for conquest. It was essentially designed for the survival of the community and distribution of grain. The most pressing threat of the time was violence and murder. Without severe restraint and punishment people would have turned to violence. People had conflicting interests; they may want the same land, the same goods or the same lovers. The denser the society, the more inescapable these conflicts.

Since Sumerians and Egyptians invented writing and accounting, they used a special person to develop a system which was the equivalent to the 'Ten Commandments'; forbidding murder, theft and adultery. The contents of the written text were determined by shamans who were the official select group (clergy). Their writings structured gods into monarchical hierarchy[122], with the national 'god' being the head of state. People were told by shamans that they should seek God's favours through worship, prayer and sacrifice. Shamans were appointed to attend to the sacrifices of animals and they acted as intermediaries between man and the 'high gods'. Sacrifices were made every day of animals, wines, food and milk. Additional special occasions called for spectacular festivity. In these feasts, 'high gods' were present. Normally these festivities took place on the day of the 'New Moon' and upon the change of seasons.

That ancient religion has left its mark on the entire Middle East and Europe in the wake of text, which only shamans have the authority to explain in a language the common people could not understand.

In ancient Egypt, religious practices were

attributed to the king of Egypt known as the Pharaoh. He acted as the intermediary between his people and the gods. He sustained his relationship with the gods through people's rituals, offerings and sacrifices. This religion was very similar to the Sumerian religion with an additional aspect of a belief in the afterlife and funerary[123] practices. The Egyptians made great efforts to ensure the survival of their souls after death. Their religious beliefs centred around the natural world and they also worshipped different animals. After the unification of Egypt in 3000 BC, the early dynastic period began. This transformed the religions, and some deities[124] rose to national importance.

Gradually the 'Sun God', known as RA, became most important. Ancient Egyptians believed that RA travelled through the underworld at night. Thus, he was supposed to be the Afterworld where RA unites with 'BA' (soul).

The Iron Age had begun. This era saw technology, such as horse-based chariots and cavalry, allowing soldiers to move faster to capture land and slaves. Empire started to emerge, obtaining control and access to trade routes as well as other cities and settlements. Emperors made laws and implemented them with brutality. During the period of 800 BCE-200 BCE, civilization was greatly affected by social change from China to the Mediterranean. From this came the spread of coinage, larger empires and new beliefs. This 'Axial Age' saw a series of male sages[125] prophets, religious reformers and philosophers from China, India, Iran, Israel and Greece would forge new directions for the world and its civilizations.

Ancient civilizations in Mesopotamia, Sumer, Akkad and Babylonia had a polytheistic[126] belief system, which means people believed in demons created by their gods. Mesopotamian religion finally declined with the spread of Iranian religion. Ancient Paganism[127] tended to focus more on duty and rituals than morality. Persians eventually conquered Mesopotamia, but religious beliefs were left unperturbed[128] by Cyrus the great (539-332 BCE). *Kurdish Yezidus* adopted several myths from Sumerian beliefs. Yezidus believed in a transcendental[129] God, who created seven great angels of which Taus Malik (peacock angel) rebelled against God. God cast Taus into fire, where he spent 40,000 years repenting and weeping. His tears put out the flames of hell and demonstrated his repentance. He reconciled with God and Taus was placed in charge of the daily affairs of the world. Taus Malik contains both good and bad, light and darkness. These beliefs still have influence on the modern world, predominantly because many biblical stories that are found in Judaism, Christianity and Islam were possibly based upon the earlier Peacock empire. The biblical religions compare Satan or Lucifer to Taus. Similarly, the creation myths such as the Garden of Eden, hell and flood all came from Yezidus. They claim that Yezidus are descended from Adam who placed his seed in a jar, creating a son and a daughter, without the help of Eve. Those two children are the Yezidus' ancestors. Conversely[130], Jews, Christians and Muslims believe that ancestry started from

Adam and Eve. Yezidus' attachment to snakes and birds represents the remnants of much earlier religions e.g. the Egyptian Pantheon[131].

During the reign of Cyrus the Great, the Zoroastrian religion flourished in Iran. This religion was based on the Avesta[132] sacred book and contained teachings of their prophet Zoroaster[133] (Zarathustra). It is one of the world's oldest religions, believing in a monotheistic creator of the world. Zoroaster was inspired by the teachings of the Avesta, their sacred text. Zoroaster taught the existence of angels, demons and the struggle of good and evil. The Avesta contains hymns, rituals and casting spells against demons. Zoroastrianism roots are found in the proto-Indo-European[134] spirituality that has also produced the religions of India. It has influenced Buddhism, Judaism Christianity and Islam. To Abrahamic religions it has given the concepts of right and wrong, monotheism[135], the concept of angels, Judgement Day and apocalypse[99] culmination[136]. The Zoroastrian central concept is that righteousness is the highest value. In the old Zoroastrian scriptures, heaven and hell are not places, but a state of mind that results from right and wrong choices.

The spread of higher religions began; Zoroastrianism, Confucianism and Buddhism were linked to the development of the Axial Age. Principal amongst this was the creation of larger militaristic territorial states, which saw an increase in the state as a powerful unit advocating the use of violence. It dissolved previous local community traditions and their moral values. The rise of confessional religions brought the ability to unify larger states that had previously existed. The Axial Age defined this as cognitive[83] change from a narrative to an analogical[137] style religion, emerging as a means of providing social and economic stability. Religion served to justify the central authority, which in turn possessed the right to collect taxes. In India, Egypt and Mesopotamia the chiefs, kings and the emperors played the roles of political and spiritual leaders. It was religion that served to provide a bond between unrelated individuals, who would otherwise be more exposed to enmity. Various religious theories were introduced through influential philosophers including monotheism[135] in Persia as well Platonism in Greece, Buddhism and Jainism in India and Confucianism and Taoism in China.

The decline of the Zoroastrian Empire started upon an increase in persecution in the 8th century, during the reign of the Umayyad Caliphs. The Umayyad Caliphs' dynasty had conquered most of the last state in 652 AD. Jizya Tax was imposed, and the official language of Persia became Arabic instead of Persian. Nowadays, India is considered to be home of the largest population of Zoroastrians. Members can only marry those within the same religion, and to add further complication, within their own caste. Consequently, the population is on the decline. In India they are known as 'Parsis' because they migrated from 'Fars' (Persia, Iran).

6.1 THE HISTORY OF RELIGION

Organised religion had served to justify the central authority. Virtually all state societies around the world had similar structure, with the chiefs, kings and the emperors playing a dual roles as politicians and spiritual leaders. In India, Mesopotamia and in China the spiritual foundation was laid simultaneously... And these are the foundations upon which humanity still subsists today.

Present-day world religions established themselves throughout Eurasia during the Middle Ages by Christianisation of the Western world; Buddhism in East Asia and the spread of Islam throughout North Africa, Middle East, central Asia and parts of Europe and India.

During the Middle Ages, Muslims conflicted with Zoroastrians during the Islamic conquest of Persia; Christians conflicted with Muslims during the Byzantine Arab wars and Crusades. Shamans conflicted with Buddhists, Taoists, Muslims and Christians during Mongol invasions and the Muslims conflicted with Hindus, Sikhs during Muslim conquest of India. The Jewish Diaspora[138] began with the Assyrian conquest and continued in a much larger scale with Babylonian conquest. Jews were also widespread throughout the Roman Empire. Caliph Omar conquered Jerusalem and the lands of Mesopotamia, Syria, Palestine and Egypt.

Abu-Rayhan Biruni (973-1048) wrote a detailed anthropology[139] of religions in the Middle East,

Mediterranean and Indian subcontinent. His methodological approach assumes that religion is created by the community that worshipped it. "That creative activity ascribed to God is projected from man."

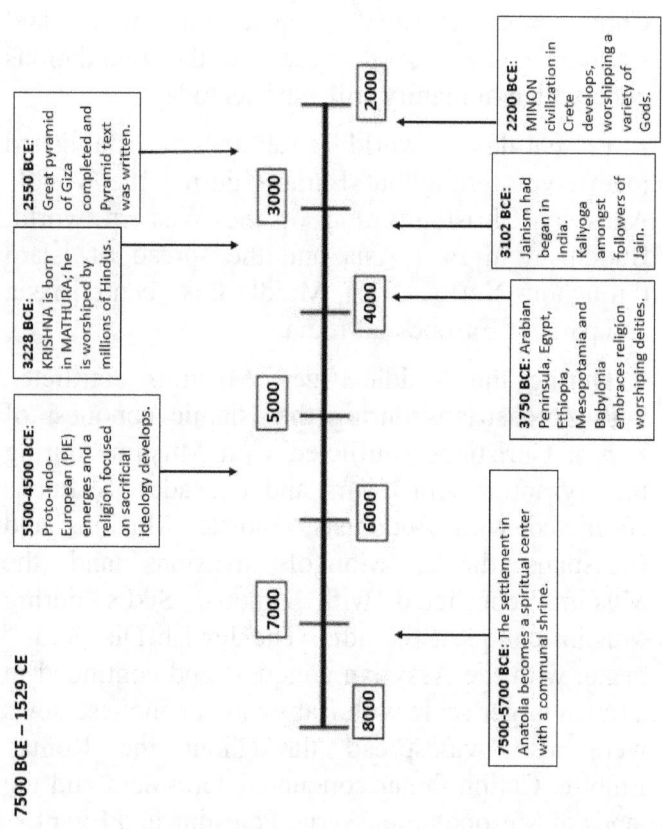

7500 BCE – 1529 CE

5500-4500 BCE: Proto-Indo-European (PIE) emerges and a religion focused on sacrificial ideology develops.

3228 BCE: KRISHNA is born in MATHURA; he is worshiped by milllions of Hindus.

2550 BCE: Great pyramid of Giza completed and Pyramid text was written.

2200 BCE: MINON civilization in Crete develops, worshipping a variety of Gods.

3102 BCE: Jainism had began in India. Kaliyoga amongst followers of Jain.

3750 BCE: Arabian Peninsula, Egypt, Ethiopia, Mesopotamia and Babylonia embraces religion worshiping deities.

7500-5700 BCE: The settlement in Anatolia becomes a spiritual center with a communal shrine.

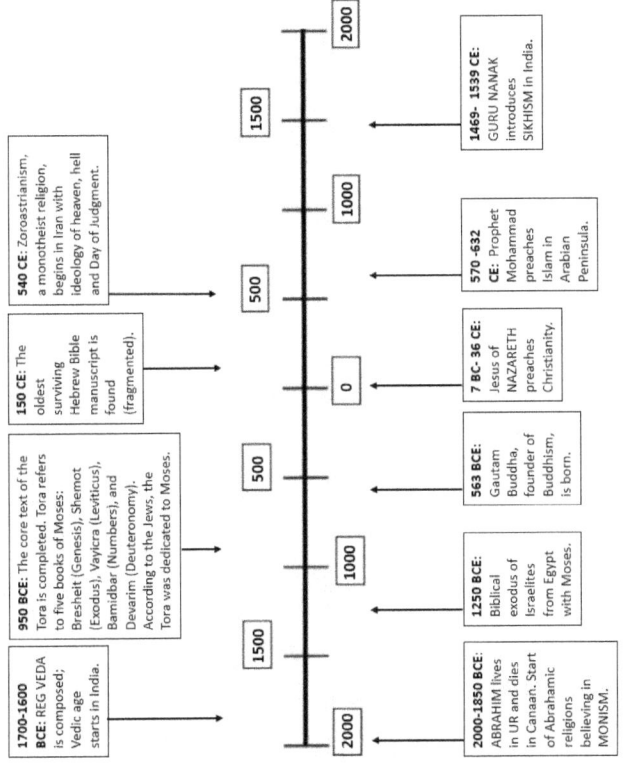

CHAPTER 7

MAJOR RELIGIONS

---

1. HINDUISM: *A body of social, cultural and religious beliefs including transmigration of souls and reincarnation[142].*

2. BUDDHISM: *The teaching that centres around the avoidance of suffering. This is inherent, and one must limit their desires and expectations.*

3. CONFUCIANISM: *A teaching of cardinal virtues, kindness, righteousness, prosperity, intelligence and faithfulness.*

4. JUDAISM: *A monotheistic[135] religion that follows the commandments[107] of the biblical book: 'The Torah'.*

5. CHRISTIANITY: *Based on the resurrection[108] and ascension of Jesus Christ, and it is a monotheistic religion.*

6. ISLAM: *A monotheistic religion that follows Mohamad's teaching through the Qur'an.*

7. SIKHISM   *There is one God. People should serve him leading a life of prayer and obedience. Eventually, the soul merges with God.*

In the previous chapter: 'Origin of Religion', we discussed the ancient polytheistic[126] and Zoroastrian religions. The following is an overview of the major contemporary practicing religions of today.

7.1 HINDUISM

Hinduism is the oldest living religion. Its roots and origin lie in the Indus Valley civilization. It has 900 million followers worldwide. Yet, Hinduism is difficult to concisely define because of the vast array[237] of practices and beliefs found within. Scholars regard Hinduism as a synthesis of various Indian cultures, with diverse roots and no single founder. Hinduism is a slowly developed thought process and in it you can see the existence of primitive religion as well as very advanced philosophical views. The religion allows hundreds of contradictory thoughts to co-exist within. On one side you see the strict morality, and on the other a tantric eroticism. Hinduism is a pantheistic religion. It equates God with the universe. Their philosophical ideas can be found in their scriptures, including Vedas, Upanishads and Bhagwat Gita.

Conceptually[140] and historically, it is closely followed by Jainism, Buddhism and Sikhism. Hindus believe that there is one true god, the supreme spirit, called **'Brahman'**. He pervades[141] the whole universe and is symbolised by the sacred syllable **'Om'**. The BBC website: 'The nature of the Hindu god' (www.bbc.co.uk) says: Hindus believe their existence in a cycle of birth, death and rebirth governed by 'Karma'. Your deeds in one life create consequences for the future. The idea of reincarnation[142] includes all ideas of good and bad, vices and values. This major ethical code is known as 'Dharma'. The particular cycle of life, death and rebirth is known as 'Samsara', with the ultimate goal of achieving liberation, when the oneness with the universal soul occurs. This achievement is known as 'Moksha'.

In the Rig Veda, we see a nomadic community just settling on the banks of the River Indus. One of the oldest scriptures, the Manusmriti, 'the code of Manu', is the history of nomad society taking roots around Indus Valley. Hinduism developed from the religions that Aryans brought to India about 3,700 years ago. Its beliefs and practices are based on the Vedas, a collection of hymns. The belief is concerned with living a good life in the hope of being rewarded in the next life.

Hindus believe in one god, the Supreme Spirit of 'Brahman' They believe that Brahman is present in every person as the eternal spirit. The Trimurti consists of three gods responsible for Creation, Preservation and Destruction of the world. Brahma the Creator, Vishnu the Preserver and Shiva the Destroyer.

Hindus believe that Lord Krishna is the complete incarnation of Lord Vishnu. There are two theories and beliefs regarding Lord Krishna and Lord Vishnu. According to Bhagwat Gita, Lord Krishna is the supreme soul and he is the source of all incarnations on Earth. Yet, in different Puranas Lord Vishnu is the supreme authority and the source of incarnations. It is said that Krishna, Vishnu and Shiva are all different manifestations[143] of a Supreme God, ONE GOD. Prominent principles in Hindu beliefs include the four goals:

Following four goals are to achieve **MOKSHA** (Liberation and Salvation):

1. *Dharma (Ethics and Duty)*
2. *Kama (Desire and Passion)*
3. *Artha (Work and prosperity)*
4. *Karma (Action and Intent)*

All life, whether human or animal is considered sacred. 'Ahimsa' forbids killing or causing harm to any living being. There are four spiritual paths which are commonly recognised within Hinduism. It is up to the individuals to choose the path of living spiritually.

The Bhagavad Gita is a Hindu sacred text originally written in Sanskrit that forms part of a larger epic 'The Mahabharata'. It is a narrative that is told through a dialogue between Arjuna, a warrior, and his charioteer Lord Krishna.

In the Bhagavad Gita, Lord Krishna comforts and advices his troubled disciple[144] Arjuna by telling him about three paths. Each he says, is a kind of yoga – a way to live in the world and at the same time maintain inner peace. They are the path of Action, the path of Devotion and the path of Knowledge. Krishna narrated about the Yoga of Action on the path of Karma Yoga. Krishna is persuasive, setting out guidelines that Karma Yoga is the ability to consciously evaluate once motivation, to act with skill and determination, and not to be attached to the outcome of the action.

In Chapters 7-12 of the Gita, he teaches the path of devotion and love (Bhakti Yoga). This path is commonly associated with those who express themselves through music, poetry, dance and other fine arts. This is a life and path of devotion. A false sense of devotion, however, may lead us in the wrong direction and therefore a Guru is needed. Lord Krishna says, "I am the same to all beings, and my love is ever the same, but those who worship me with devotion, they are in me and I am in them." He tells Arjuna that 'he who loves me shall not perish' (9:29-31). For the Path of Devotion Lord Krishna advises Arjuna that "not by study of the Vedas, nor by the austere[148] life, nor through gift giving and nor through ritual offerings can I be seen in such a way as you have seen me. He who does my work and who loves me free from attachment to all things and with love for all creations, he in truth comes to me" (11:53-55). His words ignite a deep passion in Arjuna and he gets engulfed in the wonder of the moment.

BHAKTI-YOGA:	*Followers aim to denote themselves to the love of one god or goddess such as Krishna or Kali.*
KARMA-YOGA:	*Followers perform good deeds to help others, offering all actions to God.*
GYANA-YOGA:	*The purpose to study and to understand the philosophical ideas of Hinduism. The scriptures are studied carefully under the guidance of a 'Guru'.*
DHYANA-YOGA:	*This helps to attain spiritual discipline incorporating body, heart and mind.*

YOGA is performed to purify the inner consciousness. It is believed that higher the vibrational level upon which our consciousness works, the more we can see and enter into the higher level. Whatever spiritual goal you may wish to attain can be acquired simply by engaging in devotional service to the Supreme.

The Dharma text of 'Manu' is traditionally the most authoritative of the books of the Hindu code 'Dharma Shastra'. In India it is attributed to the legendary first man and law giver – Manu. It is one of the standard books in Hindu canon and a basic text, upon which the teachers based their teachings. It tells us about the ancient code of conduct for domestic, social and religious life. The revealed scriptures comprise of two thousand six hundred and eighty-four verses divided into twelve chapters. Popular belief is that Manu had learnt the laws from Lord Brahma, the creator – so the authorship is said to be divine.

The first chapter deals with the creation of the world by the deities[124]. Chapters 2-6 recount the proper conduct of the upper castes, their initiation into the Brahman religion by sacred thread, and sin-removing ceremony. The chief duties of the householder are the choice of the wife, hospitality, offering to the gods, feast for the departed relatives and finally the duties of old age.

The seventh chapter talks about the duties of kings. The eighth chapter deals with civil and criminal proceedings and proper punishment. The ninth and tenth chapters deal with inheritance, property, divorce and lawful occupation. Chapter 11 expresses the various kinds of penance for misdeeds and the final chapter is about Karma, rebirths and salvation.

According to traditional Hindu belief, there are four stages of human life that all humans ought to go through:

1. ASHRAMAS: *A stage one goes through as student, spent in celibate[149] and controlled learning skills and knowledge under the guidance of a teacher.*

2. HOUSEHOLDS: *With various responsibilities of family life.*

3. PILGRIMAGE: *Visiting holy places, making long and difficult journeys to places in the Himalayan mountains.*

4. RETIREMENT: *Gradual detachment from the material world.*

Socially, the structure of Hinduism is based on caste classification. The Caste System is divided into four main categories:

1. BRAHMINS: *Originate from Lord Braham's 'head'. They enter professions such as priests and teachers.*

2. KSHATRIYAS: *Originate from Lord Braham's 'arms'. They are commanders,* *warriors and rulers.*

3. VAISHYAS: *Originate from Lord Braham's 'thighs'. They enter into professions such as farmers, traders and merchants.*

4. SHUDRAS: *Originate from Lord Braham's 'feet'. They can serve as labourers and perform menial jobs.*

In rural India, communities were organised on the basis of caste. The upper and lower castes lived in segregated colonies. The water wells were not shared; Brahmins will not accept food or drink from Shudras. One must marry within one's caste.

Women held a central yet ambiguous[145] role in Hindu religion. According to the laws of Manu, women are essential to the 'Dharma' of men and should find fulfilment as subordinators with subsidiary roles. The laws of Manu denigrate[150] women as unreliable, corrupt and licentious[151]. In Grihastha Ashrama, a man takes a wife and she becomes a householder. As a widow the woman loses her status in the household and becomes dependent on her sons.

Present-day scholars have criticised the work because of the rigidity of the caste system and the contemptable attitude towards women by today's standards. The Sudras, the lowest caste, were forbidden to participate in the Brahman rituals and were subject to severe punishment. Women were considered inept, inconsistent and were restrained from learning the Vedic texts. They were also forbidden to participate in some social functions. The most demeaning things for women in Hinduism were the child marriages and Devadasi system. Child marriages are still common in the Indian subcontinent regardless of religious beliefs. Yet, women in a Hindu society could achieve the highest place. Saraswathi, is the Goddess of Knowledge, Lakshmi is the Goddess of wealth, and Parvati is the Goddess of Power.

All old religions including Judaism, Christianity and Islam looked down upon women through their scriptures and mythologies. At the height of its civilization the largest empire was established by a Hindu dynasty, known as the Maurya Empire, in 250 BCE. It occupied an area of approximately five million square kilometres. It emerged as a result of the merging of states in Northern India.

Chandragupta Maurya was the founder of this empire and his policies and the laws were formulated by his ministers, which led the empire to thrive. The empire signed treaties with the Generals of Alexander the Great, when he had conquered territories in Iran and Afghanistan. During the reign of Ashoka, Chandra Gupta's grandson, the empire had conquered most of the Indian subcontinent. Ashoka was well known for embracing and re-establishing Buddhism after the conquest of Kalinga.

Hindu practices include rituals, recitations, meditations and family-oriented rites of passage.

There are annual festivals and some Hindus leave their material possessions to engage in monastic practices to achieve Moksha. Hindu worship is primarily an individual act rather than a communal[152] one. It involves making personal offerings to the deities such as water, fruit, flowers and incense as they repeat the names of their favourite gods and goddesses and they recite Mantras. The main and only aim of the soul is to develop eternal loving relation with the Supreme God. Hindus believe that devotional service ultimately brings us to the level of being at which we are able to see Lord Krishna directly.

7.2 BUDDHISM

Buddhism seeks to reach a state of 'Nirvana', following the path of Buddha, who went on a quest to seek enlightenment[82] around the 6th century BC. It is one of the oldest religions practiced today. Buddha believed that there was an afterlife and not everything ends with death. Buddhists schools vary on the exact path to liberation. We must have a moral observance and renounce craving and attachment. The ultimate goal is to overcome destructive mental states, which include ignorance and aversion[146], in order to achieve the sublime[147] state of Nirvana. Thus, with this comes the escapement of the cycle of suffering and re-birth. Scholars have a differing view about Gautama Buddha's childhood stories, claiming that some may have been invented. According to the Buddhist Sutras, Gautama was emotionally touched by the innate[81] sufferings of humans, as well as the concepts of death and rebirth. Gautama first studied under Vedic teachers learning meditation and ancient philosophies. Finding those teachings insufficient, he practiced 'Dhyana', a kind of meditation with the concept of emptiness and nothingness, under a tree in Bodh Gaya. He attained 'enlightenment[82]' and spent the rest of his life teaching the 'Dharma' that he had discovered.

Buddha taught that we expect happiness from things that are not permanent and therefore cannot attain real happiness.

'Dukkha' (Sufferings) arises when we crave and cling. As a result, we remain in the paradox[91] of life, death and rebirth. In other words, the cycle of life and death continues. This state was named 'Samsara' by Buddha, which is same as Dukkha.

A central aspect of the Buddhist theory of karma is the intent to bring out a consequence. A notable aspect of karma is a merit transfer gained by exchanging goods and services through charity to monks and nuns. In Buddhism the right 'intent' is to give up your home and adopt the life of a religious group, such that you follow a life that is reliant on begging. Survival is based on charitable donations (right intention). The moral observance should be adhered to by speaking the truth: no lying, no rude speech, no backbiting, no killing or injuring humans or animals. Don't take what is not given to you, also forbidden is adultery (right actions).

Monks beg only for feeding; they possess just the essentials to sustain life. For lay Buddhists, the correct livelihood is achieved by abstaining from money received from a source of suffering or cheating. The right mindfulness is required to achieve 'awakening'. Thus, one should always be conscious of what one is doing. Correct concentration and meditation is also required. The correct actions, speech and livelihood appear as an ethical concept of moral virtue.

As stated in 'Great Disciplines of Buddha, Monks lives and their work is their Legacy', by adopting Bodhi Bikkhu: For the general follower, these rules apply:

Abstain from killing (Ahimsa)

Abstain from lying

Abstain from sexual misconduct

Abstain from intoxicants

For monks there are eight transgressions[116] as follows:

Abstain from jewellery, perfume and entertainment

Abstain from eating at the wrong time

Abstain from sleeping on high beds

Unlike Hinduism, Buddhism does not consider

women inferior to men. Buddha emphasised women's fruitful role as a wife and a mother. Buddhist women practice Dharma now and female Sangha has been here for centuries.

7.3 JUDAISM

Abraham was the founder of monotheistic[135] religions. The name Abraham is a shortened version of a Semitic phrase meaning 'FATHER OF MANY NATIONS'. He was the direct descendant of SHEM who was Noah's son. According to the biblical books, God spoke with Abraham and gave him a central message to teach sacred values, and sacred ideas. He was also told how to relate to God, and to have faith in God. The biblical narrative commences with Abraham in Genesis chapter 12 and leads to his son Isaac, Jacob and in turn Joseph and the 12 tribes. Abraham recognised one god and launched the process that would culminate[154] in the people of Israel, who wanted to be known as Jews. Genesis 17 spells out that the circumcision of male Jews was a covenant[155] with God.

Abraham's father was a minister to the king, NIMROD. When Abraham was born, the stargazers of Nimrod told the king that his minister's new-born son would be a danger to his throne. Nimrod ordered his minister, Terah, to send his baby to be put to death. Terah, however, outwitted the king. Instead of sending his real son to the king, he sent the baby of a slave who was born on the same night. Nimrod killed the baby, believing he was safe.

Abraham along with his mother was hidden in a cave for ten years. After that they were moved to Noah's house in Ararat. There they were taught about Noah's God. At the age of 50, Abraham returned to his father's house in UR. Abraham refused the idols and angered Nimrod. He was thrown into a burning fire, but Abraham miraculously came out unharmed. Nimrod was greatly perplexed[195] and awarded Abraham with precious gifts.

Later Nimrod hatched a plot to arrest Abraham. However, Abraham had learnt about the plot and with his 300 followers, fled to Canaan. There, they would be free to worship their God. (Jewish History, "Abraham's early life" by Nissan Mindel, Kehot Publication Society).

Jews, a tiny band of nomads, were living in the upper region of the Arabian Desert; finally, they settled down in Canaan which is about 50 miles from Jerusalem. Jews were different to their neighbours, because they believed in one single, supreme, natural, transcending God. For the Egyptians, Babylonian, Syrians and others in the region, they differed from Jews in that they believed in many gods. For example, the sun god, rain god and the storm god.

Abraham's children Isaac and Jacob and their families were forced to leave Canaan for Egypt due to a famine. There they were kept in captivity as slaves. Abraham is believed to be a common ancestor of Jews, Christians and Muslims. Christians are his descendants through his son Isaac and Muslims through his son Ismail. Around the world there are almost three billion people believed to be Abraham's descendants either through his faith or blood. Moses was homed in Egypt in the period in which Jewish slaves had become a threat to the Egyptians due to their population.

Moses was born at a time when the ruling pharaohs had enslaved the Israelites in Egypt, at the time of the prophet Joseph. He was a military man and commanded the Israelites in their battles. He is also the prophet of excellence and a priestly figure involved in sacrifices. Furthermore, he is associated with miracles and helped Jews in the exodus from Egypt to escape the tyranny of the pharaohs. He guided people on the dangerous journey from Egypt to Canaan, which was their ancestral homeland. The theme of 'The Exodus' conveys the message that human beings should be free to determine the course of their own lives. It professes that they should be able to work and enjoy the rewards of that work.

Moses was a law giver, and God spoke with Moses and gave him the Ten Commandments. The Jews were instructed to live by those rules and to teach those laws to their children. Over a 40-year period in the Sinai Desert, God taught many laws to Moses and these laws were written in five books. The collection of these five books is known as Torah, which means teachings. The Torah covenant established a society which was radically different from the other societies of the time. This society became responsible for the welfare of each other and the followers were known as Jews. After Moses there had been many prophets who inspired people to act upon Torah. The words of many of these prophets are recorded in the biblical books.

According to book of Judges, the Israelites previously lived under a confederation with ad hoc[156] charismatic leaders known as 'Judges'. Biblical accounts claim that the monarchy was formed by uniting the tribes under a single state. King Samuel appointed Saul from the tribe of Benjamin as the first king. Kingdom rules of 'Saul' and 'Ishbaal' lasted for a short period. David was the king of Judah and west of Jordan. He created a strong and united monarchy, reigning from 1000-961 BCE. Hebron was the capital of Judah. David embarked on a successful military campaign against Judah and Israelite enemies. He also defeated the Philistines, thus creating a secure border for his kingdom. Israel grew from a kingdom to an empire. Its imperial borders stretched from the Red Sea to the Euphrates River and to the Arabian Desert. It encompassed about 9,300 square miles, some 24,000 square kilometres.

David was succeeded on his death by his son Solomon. Living up to his name (peace), the rule of Solomon was one in which the nation knew unprecedented peace. The united monarchy experienced prosperity and cultural development. Many public buildings were erected, including the 'first temple' in Jerusalem. David and Solomon are both portrayed in the Bible as having entered a strong alliance with the King of Tyre. In return for providing additional land to Tyre, they received several master craftsmen, skilled laborers, money, jewels, and other things. David's palace and Solomon's temple were built with the assistance of these Tyrian assets. Solomon is said to have rebuilt numerous major cities (Josephus, 93-94 AD).

Following Solomon's death in 926 BCE, tension grew between the northern part of Israel, containing 10 tribes, and its southern counterpart containing Jerusalem. This reached a boiling point when Solomon's successor, Rehoboam, dealt tactlessly with northern tribes. As a result, the Kingdom of Israel and Judah split into two separate entities. The northern kingdom contained cities such as Shechem and Samaria. The southern kingdom included Judah, which encompassed Jerusalem.

Within a short period, most of the non-Israelite provinces achieved independence. The northern kingdom of Israel existed until 722 BCE, at which point it was conquered by the Assyrian Empire. The kingdom of Judah existed until 586 BCE, and then it was conquered by the Babylonian Empire.

The focus of Judaism shifted from the sacrificial rite of the temple to the study of the Torah and its oral traditions in synagogues. Henceforth it was not the priests but the 'Rabbis' who held Judaism together.

The exile from Judah to Babylon was a major moment in the emergence of the Jewish religion (*Encyclopaedia Britannica*, 2018). For 70 years or more, they tried to retain their identity and their religion. After the return from exile in Babylon in the late 6th century, the Jewish community in Jerusalem was ruled by the temple priests. Control of the temple has remained with the descendants of Zadok, the high priest at the time of Kings David and Solomon. The aristocratic families formed an alliance with priestly families to consolidate their control of the people. The central element was known as the 'Sadducees'. They insisted on reading the Torah in the text of 'Mosaic Law' (original text). However, the scholars and the teachers were opposed to this, but did not belong to the priestly families. The Sadducees described their opponents dismissively as 'Pharisees' (they have come from Iran). The reformers eventually became associated with the piety[178] of the synagogue, where words are central in the reading of Torah, as opposed to the rituals of temple worshiping. These Pharisee priests were known as rabbis, meaning teachers.

The most important belief according to the Torah (Deuteronomy[255] 6:4) is: *"Hear O Israel, the Lord our God, The Lord is One."*

In *The World's Religions* by Huston Smith Isbn, it is stated that 'from the eighth to the sixth century BCE, Israel and Judah was overpowered by the aggressive powers of Syria, Assyria, Egypt and Babylon. The prophets found meaning on this predicament by seeing it as God's way of underscoring the demands for righteousness'.

The history of the early Jews centres on the fertile crescent and east coast of the Mediterranean Sea. It began amongst these people, as they occupied the area lying between the River Nile and Mesopotamia. According to the Hebrew Bible, Jews descended from the central people of Israel who settled in Canaan. The nomadic travels of the Hebrews centred on Habron. The children of Israel consisted of twelve tribes, each descended from one of Jacob's twelve sons – Reuven, Shimon, Levi, Yehuda, Yissachar, Zevulun, Dan, Gad, Naftali, Asher, Yoseph, Benyamin (Varner & Varner, 1987).

In 168 BC a Persian king known as Antiochus set up a statue of Zeus above the great altar of the temple in Jerusalem. Sacrifices are made to the idols here. He employed a Greek garrison in the new fortress on the site of the citadel of David. Eventually Juda of Seleucid, another Persian king, cleansed the temple in Jerusalem of Greek abominations[157] and rededicated it to one god. In 64 BC, the Roman Pompey annexed Syria and took Jerusalem. In 40 BC, Mark Antony took over the city. He tried to balance life between the Romans and Jews by appointing Herod as king, who was a Jew. Herod's kingdom gradually extended to Jordan and much of Lebanon. In that period Titus, a Roman general, destroyed the temple. Following his rule was the Roman general, Hadrian. Jews rebelled but were expelled from the surrounding region of Judea. Tax was imposed on Jews.

Around 313 BCE, Constantine issued the 'Edict of Milan' giving official recognition to Christianity.

He moved his capital to Constantinople which was the start of the Byzantine Empire. The Emperor imposed restrictions on his Jewish people; this resulted in the dispersion of the Jewish people throughout the world.

Ironically the introduction of Islam saved the Jews. The greatest concentration of Jews at that time was in Persia. Since Romans failed to conquer Persians, Jews stayed away from the influences of Christianity.

For Jews, 'Talmud' came into being in Babylon. Later, although the Muslim armies had conquered most of Africa, Asia and the Middle East, political change was partnered with the Jewish population by granting them status. The result of the Muslim conquest was to make the Jews second-class citizens. They paid extra taxes, but their lives, homes and practices of faith were protected. In those days, second-class citizens enjoyed a stable and respected life. Jews enjoyed legal and economic equality. After centuries of persecution they could move around freely, maintain contact and bring families up as Jews.

The Torah consists of five books:

1. Genesis	BERESHEET	Means in the beginning.
2. Exodus	SHEMOT	Departure from slavery in Egypt. Moses led them to promise land. Parting of Red Sea.

3. Leviticus	VAYIKRA	Taught by priesthood, moral purity. It cannot be studied independently, requires a teacher.
4. Book of Numbers	BAMIDBAR	Spoken words. ENDS with death of Moses.
5. Deuteronomy	DEVARIM	Consists of commandments[107].

The golden age of Jewish culture in Spain coincided with the Middle Ages in Europe, a period of Muslim rule throughout much of the Iberian Peninsula. During these times Jews were generally accepted in society and Jewish religion, cultural and economic life blossomed. During the classical Ottoman period (1300-1600), Jews, together with other communities of the empire enjoyed a certain level of prosperity.

During the period of European renaissance[251] and enlightenment[82], significant changes occurred within the Jewish community. Jews began in the 18th century to campaign for emancipation from restrictive laws. In the 1870s and 1880s, Jews actively discussed immigration back to Israel. In 1933, with the rise of Hitler and Nazism in Germany, the situation became worse. An extensive, organised operation, on an unprecedented scale targeted the annihilation[27] of the Jewish people. The Holocaust resulted in mass genocide of six million Jews. On May 14, 1948, the state of Israel was established by forcing Arab neighbours to allocate the area. The United States and United Kingdom forces helped Jews to achieve this goal.

7.4 CHRISTIANITY

Christianity is the most widespread of all the religions. It was founded not by the abstract principles of any society, but by a Jewish carpenter who was born in a stable and was executed as a criminal at the age of 33. He attended no college, didn't command an army and never produced a book. Jesus was born during the reign of 'Herod the Great' in Palestine and grew up in Nazareth. He was baptised by John, a dedicator prophet. For a few years Jesus had a healing career. Jesus incurred hostility from his own people and provoked[252] suspicion in Rome, which led to his crucifixion.

Christ was a charismatic teacher. His style was invitational, as opposed to telling people what to do. He invited people to see the things differently. His life consisted of humility[93], self-giving and love. His only concern was what people thought of God and God's will. People of his time loved him intensely and loved him in numbers.

We only know a few details of what happened after his crucifixion; the main point is that his followers were convinced that the death would not hold him back. His followers reported that on 'Easter Sunday', he appeared to them. Their lives were enriched with tranquillity, simplicity and happiness.

Christianity is centred on understanding the life, death and resurrection[108] of Jesus Christ. It emerged from Jerusalem and spread throughout the land. Christianity is an Abrahamic, monotheistic[135] religion. It bases itself on a belief in one God as Jesus Christ: the father, the son and the Holy spirit. Christianity emphasises the resurrection and ascension of Christ. Christianity began in the 1st century AD as a 'Jewish sect' and quickly spread throughout the Greco-Roman empire. Initially Christians were persecuted[158] by the Romans, until it became the state religion of the Roman Empire. In the Middle Ages it spread into Western Europe and Russia.

The first disciples[144] were chosen from the Jewish community. It was agreed at an 'Apostolic[160] Council' in Jerusalem that Jewish food laws and circumcision would not apply to gentiles (non-Jews). Among Jesus' disciples was Peter (original name Simon). He was regarded as the leader of the disciples. Of all the disciples, only John would die of old age. All others had to face violent deaths. According to the New Testament, Christians were subject to various persecutions from the very beginning. This involved death for being a Christian. The Great Fire of Rome was blamed on Christians by Emperor Nero. Despite all the atrocities the Christian religion continued to spread in the Mediterranean

Basin. Christianity triumphed over Paganism[127] chiefly because it improved the lives of the adherents[159] in various ways. Another factor was that Christianity combined its promise of the resurrection[108] of the dead, against the traditional Greek belief that true importability depends on the survival of the body. The pure and strict morals of the Christians, their union and their discipline gradually attracted sympathy amongst Romans. Eventually, the Roman Emperor 'Constantine the Great' was exposed to Christianity by his mother Helena. Between 324 and 330 AD Constantine built a new capital and named it 'Constantinople'. It had Christian architecture and contained churches within the city walls. Notably, there were no Pagan temples. Constantine was baptised on his death-bed.

The transition into the Middle Ages was gradual and was a localised process. Rural areas rose as a power centre while urban areas declined. Although a greater number of Christians remained in the east (Greek area), important developments were underway in the west (Latin). In the east, the church maintained its structure and character, evolving slowly. The stepwise loss of the Western Roman Empire and their dominance led to the development of Germanic Kingdoms.

In Medieval Europe a conflict between secular and religious powers began. Bishops collected revenue from estates attached to their bishopric. Bishops had no legitimate children. When a bishop died, it was the king's right to appoint a successor. So, while the king had little power in preventing noblemen from acquiring powerful domain via dynastic marriages, the king would bestow bishopric positions to members of noble families, whose friendship bishops wanted to secure. The church wanted to end this way of corruption by granting the selection of a bishop through 'Cathedral Canons'.

The Biblical Canon is a set of books Christianity regard as divinely inspired and constitutes the 'Bible'. The writings attributed to the Apostles mentions the 'Memories of the Apostles', which are known as 'Gospels'.

The religion had theological disputes that as a result, branched out into three main denominations:

1. *Roman Catholic Church*
2. *Eastern Orthodox Churches*
3. *Protestant Churches*

The Bishop of Rome claimed to be the highest and is called the Pope. The Roman Catholic Church is the world's largest religious denomination after 'Sunni Islam', and the largest Christian denomination, compromising of over half of all Christians (1.2 billion).

Catholics believe that the Pope, based in Rome, is the successor to Saint Peter, whom Christ appointed as the first head of the church. The Pope, therefore, in apostolic[160] succession, is linked to Peter and has supreme authority. Catholics are first and foremost Christians who believe that Jesus

Christ is the son of God. They also believe that the Bible is error-free and revealed the word of God.

The first actual Pope was Gregory I (590-604). This period in Europe is known as 'The Dark Ages' (500-1500).

In the Bible there are no Popes or priests to rule over the Church. The Bible never speaks of Peter being in Rome. It was Paul, not Peter, who wrote the epistle[161] to the Romans. In the New Testament, Paul wrote 100 chapters with 166 verses.

The Catholic Church teaches that a Christian soul must burn in Purgatory[162] until all sins are purged[163]. To speed up the purging process, money may be paid to a priest so that he can pray and have special masses.

Peter addresses this issue in Acts 8:20 when he says: "Thy money perish with thee, because thou hast though that the gift of God may be purchased with money." This can be interpreted as Peter's warning to followers that one cannot use wealth to buy one's way into heaven.

The Catholic Church teaches that 'Holy Mass' is a literal eating and drinking of the flesh and blood of Jesus Christ. The priest has the power to change the bread and wine into the body and blood of Christ.

In Genesis 9:4 and Leviticus 17:12:11 and acts 15:29, you will find that God absolutely forbids the drinking of blood, through the Bible. This directly contradicts some of the Catholic practices of today.

The Catholic religion is filled with all sort of

symbols and images. The Council of Trent states, "It is lawful to have images in the church and give honour and worship to them." According to God's revelation Exodus 20:4-5, "Thou shall not make any graven images or any likeness of anything that is heaven above or that is in the earth below or beneath or that is in the water under the earth. Thou shall not bow down thyself to them, nor serve them."

The catholic clergy is organised into a strict, sometimes overlapping hierarchy[122]:

POPE:	*Head of the church and based in Vatican. Pope is infallible[46] in defining matters of faith and morals.*
CARDINAL:	*Appointed by the Pope. There are 178 cardinals worldwide. They make up the college of cardinals. It is a body which advises the Pope. Upon his death, the college of cardinals elect a new Pope.*
ARCHBISHOPS:	*A bishop of a main or metropolitan diocese[164]. A cardinal can concurrently hold the title.*
BISHOP:	*Like a priest, and ordained to his station. He is a teacher of church doctrine[86]. He is a*

minister of church government.

PRIEST: *An ordained minister who can administer most of the sacraments[165] including baptism and marriage.*

DEACON: *A traditional deacon is a seminarian[166] studying for priesthood. A permanent deacon can be married and can assist a priest by performing some of the sacraments[165].*

Pope Gregory VII declared that the Pope alone could appoint or dispose of bishops. Henry IV's rejection of the decree lead to his excommunication[167] and a ducal revolt. Eventually Henry received absolution after dramatic public penance, walking barefoot in Alpine snow. This incident granted the control of selection of bishops to their cathedral canons. By the 9th century: Jutland, Denmark, Sweden and Eastern Europe including Bulgaria, Romania, the Slavs and Russia accepted Christianity. The Bible was translated into Slavonic and the old church of Slavonic was created. The missionaries, with the eastern and southern Slavs had great success because they used the native language of the people, rather than Latin as the Roman and Greek priests did.

Cultural, political and linguistic differences were often mixed with theology leading to schisms[168].

The transfer of the Roman capital to Constantinople inevitably brought mistrust and rivalry to the relations between Rome and Constantinople.

The rise of Islam, with its conquest of most of the Mediterranean coastline further intensified a physical wedge between Greek East and Latin West. The East-West schism[168] separated the Eastern Orthodox Church and Western Catholicism. The Orthodox East argued that the Papacy was taking on monarchical characteristics. They claimed that this was not in line with church traditions. The Eastern Orthodox Church, with 250 million followers, broke away from the Roman Catholic Church in 1054. Each blaming the other for the break. The Eastern Orthodox denomination remains close to Roman Catholics, but they are self-governed. The Eastern Church has no Pope. It believes that God's truth is disclosed through the conscience. They seek the consensus of Christians generally. They believe in his crucifixion and resurrection[108]. The Orthodox Church differs substantially in their way of life and worship. For example, they celebrate Christmas on January 7th. They also use the Juwan Calendar.

Because of the Ottoman conquest of the Byzantine Empire in 1453, and the fall of Constantinople, the entire orthodox communion of the Balkans became isolated from the West. This remained the case for the following 400 years. The Russian Orthodox Church was the only part of the eastern communion which remained outside of the

Ottoman Empire.

Protestants differ from the other Christian denominations. Catholics believe in the oral teachings of Jesus Christ and the Church relies on sacred traditions. Protestants believe that all that is necessary for salvation is faith in Jesus Christ and acceptance of his crucifixion as payment for their sins. Catholics believe that events described in the Bible happened first as written. Some Protestants, such as Evangelists and Fundamentalists, believe that Jesus will return at the end of time. Finally, Protestants believe that all sins are equally bad.

The Church of England is organised into two provinces, each led by an Archbishop. The Archbishop of Canterbury heads the south, and the Archbishop of York leads the north. There are 41 dioceses. Each is divided into a Parish. Each Parish is overseen by a Parish Priest.

Her Majesty the Queen is the supreme governor of the Church of England. She also has a special relationship with the Church of Scotland, which is a Free Church. In churches in England, the Queen appoints archbishops, bishops and deans of cathedrals, often upon advice from the prime minister.

Two archbishops and 24 senior bishops sit in the House of Lords, making a major contribution to the work of parliament. There are 108 bishops that are governed by a synod[169]. The general synod is elected by the clergy of each diocese. The synod considers and approves legislations for the good of the Church.

7.5 ISLAM

The Arabian Peninsula was originally the home of nomadic people, who coped with the desert climate by migrating every season. Some groups of people began to establish settlements in Mecca. It did not have a favourable climate but had the site of Ka'ba, a large cubical shrine dedicated to various deities[124]. The religion of the Arab world before the advent of Islam was an 'animistic polytheism[170]'. Most towns had their own patron god. Gradually Mecca became the centre of religious activities with 356 shrines, one for each day in the 'lunar year'. Local merchants depended heavily on pilgrims to these shrines for their livelihood. Arabian Polytheism was focused entirely on early life, and religions were not the source of morality. Blood feuds, violence and the general immorality was in abundance.

Born in Mecca, Muhammad was last in the line of Judeo-Christian Prophets. His message proclaims a new socio-religious order based on allegiance[171] to one God, Allah. The word of God was revealed to him by the angel Gabriel. The collections of these revelations were subsequently collected and codified as the Qur'an, the Muslim holy book. After the appearance of Gabriel, Mohammad (PBUH) declared that there is no deity worthy of worship except Allah and that people should put their faith in one god. He soon made enemies among the traders of Mecca. In 622 AD there was a plot to assassinate him. He escapes to Yathrib where the people welcomed Mohammad and his followers.

622 AD is argued by many scholars as the beginning of Islam, known as 'Hijra' (emigration).

Yet monotheism[135] was not unheard of amongst Arabs. There was contact with Zoroastrianism, which was the official religion of Persia. It was a dualistic religion with beliefs in Heaven and Hell as well as the Final Judgment. In addition, both Judaism and Christianity were present in the Arabic Peninsula. In Yathrib, the Jewish population was especially influential. However, monotheism was not a popular creed[213] with those whose livelihood depended on idols.

Muhammad's message was unpopular among the leaders of Mecca, and they forced Muhammad and his followers to emigrate to Yathrib (Medina). This emigration in 622 is known as 'Hijra', which marks the beginning of the Muslim calendar. Mecca was a prosperous city, its wealth and influence was based on the caravan trade, as well as Kaaba, a shrine and a place of pilgrimage housing the Pagan deities[124]. Mohammad's message continued to attract followers and within a few years Mecca embraced Islam. There had been authentic prophets of God before Mohammad, but he was the final prophet. Hence known as 'The Seal of the Prophets', no valid prophet will follow him.

Muhammad, upon his return from Medina to Mecca. assumed a different role. He became a masterful politician and administrator. History depicts[172] his administration as an ideal blend of justice and mercy. He welded the five conflicting and heterogeneous[173] tribes of the city, three of which were Jewish, into an orderly confederation. His reputation spread, and people began to flock from every part of Arabia to see him. He was God's "standing miracle", unschooled and unlettered. He was unable to read and write. Thus, how could he have produced a book that provides the ground plan of all knowledge? At the same time the Qur'anic writings were grammatically perfect and poetic. It defies belief, it is beyond comprehension[84]. The Qur'an is divided into 114 surahs (chapters), which apart from the first Short surah (part of a Muslim's daily prayer) are arranged in order of decreasing length. Surah Two has 286 verses, the third Surah has 200, down to Surah 114, which has only six verses.

As stated by Huston Smith in *The World's Religions*: "The Quran does not ground its theology in dramatic narrative as the Indians epics do, nor in historical ones as do the Hebrew Scriptures; nor is God revealed in human form as the Gospels and the Bhagavad-Gita" (Smith, 1958).

Relations with Mecca deteriorated to the point of pitch battles. However, diplomacy won the day, and Mohammad persuaded Meccans to allow his followers in 629 AD to undertake a pilgrimage to of Ka'ba (Hajj). Local citizens were impressed by the discipline and the unity of Muslims. They departed peacefully after the agreed three days. In the following year the Meccans broke the truce, provoking Muslims to march on the city. Muslims took control of Mecca without resistance and the inhabitants accepted Islam. Mohammad swept the idols out of Ka'ba. Since the migration to Yathrib,

Mohammad commanded 27 battles and fought in nine of them. These battles included: Badr, Uhad, Al-Khandaq, Banu-Quresh, Banu-Al-Mustalaq, Khyber, Conquest of Mecca, Hunyan and Al-Taif. These battles were against the injustice and torture of Muslims, who were permitted to fight to defend themselves. During the battles, Muslims were forbidden from murdering woman, children and people who did not participate in wars. After the battle of Mecca, Mohammad only lived for two more years.

After the death of Mohammad, Abu Bakr was elected by Muslims to lead them and was known as the first 'Caliph'. It should be noted that there was no clear successor to Mohammad among his followers. The likely candidates were Abu Bakr (the father of Mohammad's wife AISHA) and Ali (cousin and son in law of Mohammad). Abu Bakr was elected while Ali was away. This consequently paved the way for dispute. He only ruled for two years but within that brief period, Muslim armies had begun their astonishing expansion, subduing the whole of Arabia and striking as far north as Palestine.

The basic theological concepts of Islam are virtually identical to those of Judaism and Christianity. Its message was to remove idols from the religious scene and focus on the divine as a single invisible God for everyone. In surah 6:12, 59, "Is he not closer than the vein of your neck? You need not raise your voice, for he knows the secret whisper, and what is yet more hidden...He knows what is in the land and in the sea; no leaf falls but he knows it."

There are five Pillars of Islam:

- *The first of the five Pillars is the confession of faith (Shahadah). There is one god and Mohammad is his prophet.*

- *The second Pillar of Islam is the Canonical Prayer (Salah). This includes washing, purifying the body and showing gratitude and submission to Allah.*

- *The third Pillar of Islam is charity; material things are important, but some people have more than others. Those who have more should help lift the burden on those who are less fortunate. The Qur'an sets this tax to be 2.5%, not just of one's income but one's holdings on a yearly basis.*

- *The fourth Pillar of Islam is the observance of Ramadan, that is, to fast during the period of Ramadan. It teaches self-discipline as well as making one feel empathy and compassion[96] for those who are hungry.*

- *Islam's fifth Pillar is pilgrimage. This should be done once during his or her lifetime if one is physically and economically able to do it. It removes all barriers between rich and poor such that they can share a loyalty for the transcendent[38] and omnipotent[40].*

Abu Bakr was succeeded in 634 AD by Omer. He ruled for ten years. During his reign, largely due to superb leadership, his rule was extended to Iraq, Iran, Syria, Azerbaijan, Armenia, Georgia, Afghanistan, Turkmenistan and the Byzantines. The Byzantines had lost more than 75% of their territory and the Sassanid[174] Empire ceased to exist. Omer was an incredibly powerful and influential Caliph. He had a strict and autocratic[175] nature. Omar instilled[176] a policy of religious tolerance, which was received gratefully by Jews and Christians. He also instituted two taxes: 'Kharaj' from land owners and 'Jizya' paid by non-Muslims for their protection. His victory over the Persians led to them becoming fiercely nationalistic people. Omer was assassinated in 644 AD by a Persian Christian. At the time of Omer's death, the Muslim empire was only second to the Chinese Empire in size.

'Uthman, a member of the Umayyad family, was chosen as Omer's successor, leaving Ali's supporters once again disappointed and angry. Uthman served as the third Caliph from 644-654 AD. He expanded the Muslim rule to the west of Egypt, conquering Cyprus in 649 AD. He appointed the fellow members of the Umayyad family to administrative positions. They depleted the treasury with lavish spending habits and lacked a stable financial plan. Uthman collated all of the Qur'anic texts and completed the first handwritten Qur'an. This significantly reduced disagreements over the doctrine[86]. In 656 AD, 'Uthman was assassinated by a group of Egyptians. Discontent increased, and a civil war erupted. There was Muslim infighting to assume leadership and power.

After the assassination of 'Uthman, rebels had to elect a new Caliph. They asked Ali to accept the Caliphate. Ali was the Caliph between 656-661 AD. It was the toughest period in Muslim history, coinciding with the civil war. He reigned over the Empire as it was extended from Central Asia to North Africa. Many Muslims take a negative view of his governance, citing justice and tolerance on one side and harshness on the other side.

Ali's humility[93], austerity[177], piety[178], deep knowledge of the Qur'an and nobility gave him great distinction among the companions of the prophets. Ali was also a great scholar of Arabic literature.

During his reign the situation in Hijaz (Mecca & Medina) became so troublesome that Ali moved the capital to Iraq. Muawiya (Governor of Syria) openly rebelled against Ali and a fierce battle was fought between their armies. The battle was inconclusive, and Ali had to accept the de facto[179] government of Muawiya in Syria.

A fanatical group called 'Kharijites' consisting of people who had broken away from Ali due to his compromise with Muawiya claimed that Ali was not worthy of rule. They assassinated Ali in 661 AD outside a mosque in Kufa. Ali was the last Sunni Rashidun Caliph and the first Shia Imam. After the death of Ali, people were divided regarding their view towards him. Gradually Ali admirers turned into the majority, due to the misrule and repression[180] of Umayyad. After Ali's death, the 'Shias' of Iraq declared Ali's eldest son Hassan successor of Ali. However, he did not want further bloodshed and abdicated, being replaced by Muawiyah. Muawiyah died at 45 and was succeeded by his son Yazid, but Hassan's brother Hussain refused to accept Yazid's leadership.

In the Umayyad Caliphate (661-750), Muawiya was the leader of the struggle against Ali and his supporters. He established himself after Ali's death as undisputed Caliph. Damascus became the capital. Muawiya was the member of one of the most prominent families of Mecca. However, against considerable opposition, he still managed to establish a key new principle: that the role of the Caliph shall be hereditary[181] rather than elected. For the next century it would be passed on within the family.

From the 7th century, after the death of Ali, opponents of the new Umayyad dynasty promoted the claim of Ali's two sons, Hassan and Hussain; these were the Shias. Their party became known as Shiite Ali. However, after the death of the brothers, their political cause crumbled. Hassan died in 669. Hussain, subsequently the most Holy Shiite martyr, was killed in the battle of Karbala in 680 AD. Their faction[182] is in lasting religious disagreement with the Islam of Caliphs.

The main group under the Caliphate became known as Sunni (orthodox rule) and the new schismatic[183] sect acquired the name Shiites.

By the 8th century, devout Muslims struggled to retain the purity and mystical fervor[184] of the early years of their religion. During the explosive first century of the Arab expansion, the relationship between Arabs and Muslim changed. Originally, the two were inseparable. The Muslim armies were made up entirely of Arab tribesmen. However, now only Arabs could be recognised as Muslims. When the Muslim expansion had reached its peak, there were not enough Arabs to provide troops.

Muslims from other groups were fighting alongside the Arabs. For example: Berbers in the west and the Persians in the east. Inevitably, resentment built between the two. Non-Arabs often felt that they were treated as second-class Muslims. The various tensions and difficulties associated with controlling a vast new Empire resulted in rebellion against Umayyad Caliph.

7.5a THE ABBASID

In 747 AD, a revolt in Persia was headed by descendants of Al-Abbas, an uncle to the prophet Mohammad. Their new Caliphate, established in 750 AD, would be known as 'Abbasid Muslims', present in the east. They were Persian, and their sophistication was beginning to divert Muslim culture from its simple Arab origins. Abbasid forces reached and captured Damascus in 750 AD. Consequently, Abu Abbas was proclaimed as the

Caliph. Male members of the Umayyad family were hunted down and killed. The new capital of Persia became Bagdad. Here, Islam outgrew its roots and became an international religion. The Caliphate was known for its prosperity during the reign of Harun-Al-Rashid.

By the end of the 8th century, a distinctive Arab civilization was beginning to emerge in separate regions. Bagdad was in the east and Cordoba was located in the west. The shared characteristics of the eastern and western Empires was Islam. Its tolerance allowed Jews and Christians to play a full part in their communities. Some common characteristics from Spain to Mesopotamia included the Arabic language and a commitment to Islam.

An aggressive Turkish dynasty would later win power in Ghazni (south west of Kabul). Their ruler, Mahmud, raided Peshawar and various parts of India. Expeditions into India became regular policy. Many of the soldiers involved in such ventures were known for their affiliation with looting and their violent acquisition of property. Eventually the Moghuls settled, laying the foundations for what would become the most spectacular Empire of India.

During the 15th and 16th centuries, Islam spread. Areas that it reached included the Malay Peninsula and the islands of Sumatra and Java, Indonesia. By the mid-16th century, Islam commanded regions from the Atlantic coast all the way across to India. There were three powerful neighbouring Muslim Empires:

- *The Ottoman Empire, with its capital in Istanbul.*

- *The Safavid dynasty of Persia. It was passionately committed to the doctrines[86] of Shiite, in opposition with Sunnis of Turkestan.*

- *The Moghul Empire although occupying greater parts of India as a ruling class, were in the minority.*

During the 18th and 19th centuries, Iran was shaken by internal conflict and there was a clear external threat from Europe challenging the Ottoman Empire. Added to this, the Moghuls had given way to the British Raj.

7.5b FALL OF MUSLIM EMPIRE

The Islamic civilization kept on flourishing despite the vices that occurred with misuse the power by the elite. This corruption was particularly evident where 'women' and accumulation of wealth were concerned. However, the freedom of intellectual pursuits facilitated the development of great centres of learning in Cordoba, Baghdad and Cairo.

The 'Genghis Khan Conquest' left large areas of the Islamic region of central Asia depopulated, as every city and town that resisted Genghis Khan was met with destruction. The 'Mongols' conquered either by force or through voluntary submission.

The areas today know as Iran, Iraq, Syria, Palestine and Turkey were torn apart. In 1258, Hulagu Khan destroyed numerous libraries and the works of scholars including the House of Wisdom in Bagdad. He conquered much of Western Asia. The Mongols looted, as well as destroyed mosques, palaces and Caliphs. The treasury was also plundered. Khan's soldiers totalled around 100,000 in Bagdad, yet 200,000 Muslim soldiers were killed. The destruction was considered 'The End of the Islamic Golden Age'. The Caliph Al-Mustasim was locked in his palace tower, with nothing to eat but gold, in order to illustrate the hypocrisy that engulfed his regime. This is depicted[172] in *Livre des merveilles du monde* by Marie-Thérèse Gousset.

Gradually Islamic scholars come to the view that 'Islamic civilization' had reached its 'apogee[185]' and all the 'Ijtihad' had been accomplished. They believed that this destruction at the hands of the Mongols was the retribution[8] of Allah.

CHAPTER 8

ISLAM: THE REASONS FOR

DECAY

Civilizations rise and decay, empires rise and fall. The reasons for the rise and fall of empires are less complex than the rise and fall of a civilization. One clear reason is that whilst empires require military might and arms, civilizations require the power of ideas. These ideas are nurtured by the people who work towards the betterment of society. When ideas must be forced upon the people, the system of justice suffers.

The most prevalent[186] diagnoses and remedies for the decay of Islamic civilization fall into two categories.

1.   *The most popular view is that the Muslims have veered[187] away from the teachings of Islam. The*

remedy is that we must become good Muslims, then we will regain momentum and be able to revive the grandeur of the past.

2. *The second conventional view is that the colonization of Muslim land by western powers has stifled[188] Islamic civilization. This can be remedied by getting away from materialism and supporting Islamic education. As such, the beneficiaries of this will be able to contribute to intellectual growth and Islam can become great again.*

Upon closer survey of history, it appears that deviation from the teachings of Islam started immediately after the death of the prophet Mohammad. As soon as Mohammad died, various tribes rebelled against the selection of a Caliph. It was the subtle handling of Abu Bakr that brought these rebellious tribes back together, requiring strenuous persuasion. Caliph Omer, after ten years of rule, was assassinated. He was accused of mismanagement. The reign of the fourth Caliph, Ali, resulted in a civil war. Ali was assassinated by the 'Kharijits'. Mauwiah changed the protocol for the selection of a Caliph, from the Islamic democratic way into a hereditary[181] position. His son Yazid mercilessly killed Hussain and his entire family (barring one child) with the aim being to ensure that power stayed with him.

So why did Islam spread into huge areas so quickly, despite all these deficiencies? One could argue that Islam was interpreted by the conquered people as an egalitarian[189] religion. The lives, beliefs and the properties of the people were protected. The survival and flourishing of communities of Christians and Jewish in the heartland of Islam, and the majority of Hindus in India is testament to the tolerance of Muslims of the time.

Albert Hourani states, "The question of why the people convert to Islam has always generated intense feelings. Earlier generations of European Scholars believed that conversion to Islam was made at the point of the sword, and that the conquered people were given the choice of conversion by force was in fact rare. Muslim conquerors ordinally wished to dominate rather than convert. In most cases worldly and spiritual motives blended together."

Moreover, conversion to Islam did not necessarily imply a complete turning from an old to a totally new life. Whilst conversion entailed the acceptance of new religious beliefs and followed the rituals of praying and fasting, most converts retained a deep attachment to the culture and customs of communities from which they belonged. The result of this can be seen in the diversity of Muslim societies today with varying manifestation[143] and practices of Islam. The scholars of Islam today focus more on rituals and prayers, rather than morality and values as were preached by the Prophet Mohammad and his companions.

8.1 QUR'AN'S INTERPRETATION

Islam is the complete and universal version of a faith that has been revealed before through many Prophets including Abraham, Moses and Jesus. Muslims believe in all the prophets and scriptures such as the Torah and the Bible. One difference is that they believe that, over this period, some previous messages and revelations have been partially changed or corrupted. Muslims consider the Qur'an to be the word of God as revealed to the Prophet Muhammad. Most Muslims also follow the teachings and practices of 'Sunnah' as recorded in traditional accounts. There has never been any question about the ontological status of the Qur'an as a divine speech of sacred text. However, the problem is not in the text itself, but in how we interpret this sacred text. Since the Qur'an was addressing the 7th-century Arabs, it is written in a language and in a way relevant to those people's lives. The Qur'an universally defines a system of morality that is required to attain salvation. The Qur'an also states vices clearly.

It is possible to distinguish between what is universal and what are historical events in the Qur'an. For example: the provision of polygamy[190] is only acceptable with female orphans and crucially, it is 'conditional' upon the betterment of their life. At this point, it is worth mentioning that there was no social security or structure in place for orphans. Thus, conditional polygamy was offered as a solution. However, some Muslims have interpreted this to mean that they can have numerous different wives to serve their own sexual desires. We must look at historical reasons, instead of blindly and ritualistically imitating things when it suits us. Another key issue is the necessity to understand and interpret the Qur'an, such that it delivers key moral qualities in a 21st-century society.

It is the right of every Muslim to read and interpret the Qur'an for themselves. Why is it left with mullahs as their linguists and scholars to elaborate[98] on it? In many places and communities, the people who define religious knowledge are primarily men. Consequently, this provides an opportunity for interpretations to be 'male' oriented and anti-patriarchal[191].

The Qur'an should provide constructive contributions to gender equality. Furthermore, many Islamic scholars provided engineered interpretations of the Qur'an, designed to suit the Muslim Monarchs. These authorities of their time used these interpretations to give divine legitimacy to their tyrannical regimes. Islamic Sharia law is a product of human thinking and is certainly susceptible[192] to re-thinking.

Originally, the text of the Qur'an was written without diacritical[193] point (which helps distinguish some letters from others). However, early in history diacritical points were clearly added. Since Muslims believe that the word of the Qur'an is revealed by Allah, the act of reading and reciting the Qur'an is believed to be a means of receiving blessing from Allah. It is not uncommon to recite the Qur'an without having any understanding of the words. Even those who cannot read, believe they can benefit from hearing it being recited. The Qur'an, in its purest and most original form, was written in classical Arabic. Modern Arabic differs in grammar, although there is no difference in vocabulary and spelling. There are two types of verses in the Qur'an.

1. *Verses with basic meanings.*
2. *Verses with allegorical[194] meanings.*

Those who do not differentiate between the 'basic' and the 'allegorical' verses will surely get confused when they find two conflicting messages. Therefore, we must know the difference between verses, and allegorical meanings must be sought in the light of basic verses, which are normally very clear.

One could be perplexed[195] by the fact that there are so many religions. Are these religions contradictory or complimentary? One could argue that they are all spreading the same moral values. Perhaps God does speak the same truth to people all over the world. The question is, when God sends messages, what do people hear? Two people can listen to the same story and come away with different meanings. We hear what we need to hear, in order to face our own particular challenges. Thus, interpretation is what is important. No matter what religion we look at, the people who believe in it, think they are the ones who have the real truth about God and all other religions are a little off, or a lot off.

8.2 ISLAMIC FUNDAMENTALISM

Western interpretation of certain Islamic groups, described as fundamentalists, stems from their use of the term in the context of Christian fundamentalism[196]. The term was used in North America, by the protestant Christians in the late 19th century, to describe the conservatives, due to their opposition to liberal interpretations of the Bible, the divinity[105] of Jesus and the second coming of Christ. Although the movement did not prevail after its initial hype, a resurgence[197] of the term 'fundamentalism' came about in the 1970s with a decline in the standards of **morality** and the rise of the groups in Islamic states that were prepared to take arms.

Muslims believe that through the divine revelations of the Qur'an, the angel Gabriel corrected errors which may have occurred over long periods, in Jewish and Christian scriptures. According to Muslim faith, followers must believe in the Torah and the Bible as well as the Qur'an. Traditionally, Muslims, at the height of their empire tolerated Jews, Christians, Hindus, Zoroastrians and Shamanists. History cannot show any suggestion of genocide[198] throughout the Muslim empire. 'Jihad' meant the personal and individual struggle against basic human instincts and selfish ambitions.

The Islam that I have lived with and respected as part of my personal religious faith has now shown a wild side as a reactionary defense against the modern world. Islam, the most tolerant and most accommodating of religions now contains, paradoxically[91], people with little knowledge and misguided faith. These people want to prove their piety[178] and righteousness with destructive deeds, condemnation and violence. Since 9/11, the crusade seems to have re-emerged, enmity between Islam and the West has grown. Since Russia's demise, the USA needed an enemy to target. The USA, and its Central Intelligence Agency (CIA) in particular, was responsible for supporting Arab volunteers who came to Afghanistan to wage war (Jihad) against Soviets. American dollars and arms poured in to teach the Taliban (Mujahidins) to fight the infidels and consequently gain access to 'Heaven'. They were trained by the USA and Pakistan. Meanwhile the 'Muslim Brotherhood', a radical Islamist organisation, carried on providing volunteers in exchange for money given by the USA through Saudi Arabia. The US supported war against the Soviet Union and Saudi Arabia supported the Taliban with money.

When these 'Mujahidins', as Margaret Thatcher referred to them, turned against the USA, they were clueless about how to tackle this monster which was created by them. The Bush administrations, along with Tony Blair, were misled to invade Iraq. They were clearly isolated from the reality. They did not give any thought to what would happen after this invasion of Iraq. They repeated some mistakes with Libya, Yemen and Syria. Today there are two million Muslims in these countries who have lost their lives and some 59.5 million refugees with no home. Yet, Bush and Blair have not been tried for war crimes. Perhaps the region would have stayed peaceful with Saddam and Gadhafi still there. The Libyans, after the US invasion, are still engaged in conflict to seek control. Ironically ISIS arm themselves with US-made military hardware to wage so-called 'Jihad'. Arms factories in the USA and Russia are mass producing and shipping those arms to rebel groups with the new world order. The media is controlled by a few masters who instruct them to give air to Islamophobia.

Today, the Muslim world is corrupt, stuck in superstition, divisive and strewn[199] with religious rituals that bear little significance in today's world. Despite the diversity of the Muslim world, Muslims are forbidden from translating the Qur'an and scriptures into a language that they can read and understand. The lack of critical thinking, and an education which is irrelevant to the society they live in, has made it all go wrong, yet the spirit of Islam is professed in the Qur'an undisputedly. It is only the Muslims that can understand and follow its simple messages, rather than the books of Hadith, which often lack authentication[200].

Muslim fundamentalism[196] is a serious threat because it represents terrorism, religious fanaticism and the exploitation of social and economic structure. The act of terrorism by Muslims, however, must not be confused with fundamentalism, for they do not represent the ideology of the fundamentalism.

Over the centuries and in its historic context, Islamisation is a consequence of Western domination over Muslim land in the Middle East and beyond, during the 19th and 20th centuries. After World War I, most Muslims had to succumb[201] to the dictation of non-Muslim rule. With the withdrawal of colonial powers after World War II, Muslims found freedom in expressing themselves politically. Thus, the resurgence[197] of Islam must be viewed against a backdrop of a long period of subjugation[202] resulting in a growing resentment towards the moral misconduct and sexual laxness of the West.

Islamic countries, following the days of colonisation, must find a balance between a desire for the ideal Islamic states and adopting to the modern world. For example, Muslims are forbidden from taking usury[253] from the bank, and therefore, are not allowed to take up loans and mortgages. As a consequence, Islamic banks have adopted a system to offer loans at exaggerated prices, such that interest does not have to be paid. Effectively, this hypocrisy allows these banks to make money whilst following the technicalities of Sharia law.

In the post-World War II era, Islamic Fundamentalism became a totally reactionary and counter-revolutionary phenomena. In the 1950s, 60s and 70s, there were strong communist movements infiltrating the Islamic world. Syria, Yemen, Somalia, Ethiopia and some other Islamic countries experienced political coups and the overthrow of feudalist regimes. One of these leaders was Colonel Jamal Abdul Nasser, who became the President of Egypt. One of his policies was the nationalisation of the Suez Canal. This involved insisting that anyone wishing to use this trading route would have to pay heavy fees, directly provoking the Western world. Particularly frustrated by this were the British and French. In the 1956, the Suez war resulted in Britain and France suffering a humiliating defeat. Shockwaves were sent through Washington and other centres of imperialist power.

The Muslim Brotherhood of Egypt and Jamaat-e-Islami of Pakistan have constantly sought to promote Islam through peaceful means. Muslims themselves do not recognise the term 'fundamentalism'. There has always been a common theology for Islam, and the Qur'an has never been questioned as the authentic word of God. It returns to the same question; how is the Qur'an interpreted by its representatives (Mullahs and Ulmas)?

Islamist groups, including Hamas and Hezbollah in Palestine, participated in political and democratic processes as well as taking up arms. The chief objective of these parties was to facilitate the abolishment of the state of Israel. Conversely[130], radical Islamist organisations such as Islamic Jihad, Al-Qaeda and the Taliban rejected democracy in favour of violence. This brutality was used as a tool, enabling the implementation of Sharia Law into society. Islamic fundamentalism is widespread, with a foothold in large parts of Asia and Africa.

The Iranian revolution in 1979 carried out by Ayatollah Khomeini had turned Iran into an Islamic republic. In that republic, women were required to be veiled and clothed heavily, alcohol and western-style entertainment was banned. Although his rule was authoritarian, it retained some elements of democracy. However, the compulsory veiling, known as Hijab, was viewed by the west as the lowering of the status of women and it has often been criticised as a symbol of resistance to modernity.

In 1978, the invasion of Russia into Afghanistan and the virtual takeover in the following ten years prompted the USA, with the help of Saudi Arabia and Pakistan, to build an army that would combat the Russian troops in Afghanistan. The religion was used as a tool, and Pakistan was used to promote 'Holy Jihad' against Russia via the Taliban who were taught about Islam in Madrassas and were trained by the Pakistani forces along with British generals. The funding and arms came from Saudi Arabia and the USA respectively. The Taliban belonged to the poorest of the communities in Pakistan. They had never been educated in proper schools and were brainwashed by the so-called Mullahs in Madrassas. Those teaching were hardly related to Islam itself. These Taliban were then sent over to Afghanistan under the command of American generals to fight against communist ideologies. After the Russian forces evacuated Afghanistan, the whole of the government system was left to its own devices where the Taliban, out of greed and misguidance, turned against the appointed civil government as well as the American government. These Talibans (freedom fighters) were then named 'fundamentalists' and 'terrorists'.

The events responsible for the international outcry against Muslim terrorists including suicide bombings and Jihad (holy war) waged against the infidels are attempts by Islamic purists to purge society from non-believers. These events and ideologies of purists can be traced back to Saudi Arabia: in particular, Osama Bin Laden and his group. There is now a realisation that all these atrocities had been carried out with the collaboration of the USA, in terms of funding of cash weaponry to that group.

For the last 50 years or so, Islam has been under social scrutiny following the attack on the World Trade Centre. The term 'Islamic Fundamentalism' can be applied to Muslims with an agenda for social reform that seeks to redress western influence in its culture. The term, however, is meaningless and at best shows misguided understanding of Islam and its underlying principles.

Upon tracking historical evidence, some actions of the CIA must be questioned. It can be argued that their planting of opium (from which heroin is extracted), for the lining of the pockets of the warlords, led to the drug epidemic that spreads from this region today. This feeling of immorality that enshrouds the US only develops further upon the consideration that heavy arms were provided to these regimes by the Americans. The rationale was to provide them with military power as a deterrent to the Eastern communist regime. Fast forward to today, and the drug policy of the CIA is having a disastrous impact on youth around the world.

This hypocrisy is that Saudi Arabia, who is totally dependent on Americans for its defense and arms, has never been criticised by the USA for not abiding by human rights laws. The USA seems determined to politically attack regions such as Afghanistan and Yemen, yet 'Islamic State' is funded by Saudi Arabia. Why have the Saudi Arabians or Americans not been held accountable for this?

The reality is that in Islamic countries, large sections of the youth population are deprived, frustrated and bewildered through unemployment and lack of proper education. This raw youth is shocked by a significant section of the ruling class that has been mercilessly plundering the state and the society. This elite class have control over the judges, police, the media and representatives of the people through a so-called democracy. Shocked by the social and cultural conditions in which they live, the youth try to find solace[203] and peace for their soul and mind through Madrassas. Islamic organisations in turn exploit this youth. In these countries, the leaders have failed to provide healthcare, education and jobs; the youth is ready to be exploited. In addition, religious schools (Madrassas) have been built. Their purpose was to provide religious education for the youth, but some of them have deviated from this core principle. They have instead trained fanatics from a very young age to become raw fodder[204] for religious frenzy[205]. These Madrassas have become the only means of providing any kind of education for children that belong to poor families. These poor families cannot feed, clothe or educate their children. Thus, people in this position can either let their children suffer the horrors of child labour or send them to these Madrassas. Here, they are often grown into hysterical fanatics, prepared to take human lives for causes they don't even understand.

Islamic fundamentalism[196] is a reactionary phenomenon, representing a society that has stagnated due to feudalism and racism. The responsibility rests on a society that is corrupt to its core. Religion is used to provide some sort of soothing effect on people that belong to a lower stratification[65] of society. The majority of Muslims read and perform rituals in Arabic, which they don't understand. The true values of Islam are hardly preached or talked about. Non-Arabs don't understand the Qur'an; even Arabs struggle to understand the Holy Qur'an because Quranic Arabic is different to modern Arabic (as discussed earlier). To get rid of fundamentalism we must eradicate[74] the basic causes and systems upon which society stands. These causes include:

- *Poverty*

- *Lack of education*

- *Corruption.*

CHAPTER 9

CULTURE AND SOCIETY

Before I delve[206] into what forms a society and the cultures that it embraces, it is important to understand what these terms mean.

CULTURE: *The norms, rules and expectations from the people of a society. This is to guide us towards a certain behaviour.*

SOCIETY: *The aggregate of people living together in an orderly community. Hence, culture exists within the framework of a society, and as such, creates the dynamics[207] of day-to-day life that occur within the community.*

Culture consists of the beliefs, behaviours, morals and other practices that are shared within group members. Cultural transmission from generation to generation seems to depend on copying. The human tendency to imitate far exceeds other species. However, explaining a behavioural pattern alone cannot lead to conditioning. Therefore, the explanation of something alone does not have the same reinforcement value as learning from watching those around you. One must lead by example. Culture is historically transmitted from one member to another through teaching. Through culture, people and groups define themselves such that they conform[90] to a society. The cultural shared values become the 'norms' of the society. Norms can be defined as the rules or behaviours in specific situations. Those rules identify what should be judged as 'good' or 'bad'. Value and norms may differ significantly amongst cultures.

The most basic universal values are: human health, justice, education and equality. For the protection of these human values, acknowledgement and development are required. To achieve that, institutions are formed. These institutions refer to a cluster of rules and the cultural meanings associated with specific social activities. Common institutions are: religion, family, health, education and work. The cultural bonds may be ethnic or racial, and are based on shared beliefs, values and activities.

The belief that culture is symbolically coded can thus be taught from one person to another. Cultures, although bound are also susceptible[192] to change. Cultures are both predisposed[208] to change and resistant to it. Resistance could come from habit, religion, and the integration and independence of cultural traits[209]. Cultural change can have many causes, including the environment, inventions and contact with other cultures via immigration. Contact between cultures can also result in replacement of the traits of one culture with those of the other. Society and culture co-exist because humans have social relations and meanings. Culture can and does change over time as society's norms and habits change, but the members of the society govern that change. Accordingly, individual members of the society have a level of control over this culture.

Society provides us with a system and a platform to work together, for the betterment of the world. With the collective efforts of society, we are able to improve our living and social conditions. We have electricity, because someone invented the idea to generate electricity and so we all benefit. We are advancing due to collective social efforts. Without society, we are just *Homo sapiens*[47].

Regarding the status and roles of the people in a society, most people associate status with the prestige of the person's lifestyle, education or vocation[210]. According to sociologists, status describes the position a person occupies in a particular setting. We all occupy several statuses and play roles that may be associated with them. A 'role' is a set of norms, values, behaviour and personality characteristics attached to a status. An individual may occupy the status of a teacher, employee, club president, husband and a father all at the same time.

Society decides what is considered an appropriate role for a specific status. For example, every society has a 'mother' status. However, some societies consider it inappropriate for a mother to assume a role of authority in the family. Other societies ascribe lots of power to the status of the mother. In Eastern societies, students are expected to be completely obedient to their teachers. In American society, the role of the student is different in that it involves questioning the teacher and even challenging the teacher's statement.

9.1 SOCIAL STRATIFICATION

Social stratification[65] is the system by which a society ranks categories of people in a hierarchy[122]. In the USA, some groups have greater status through power and wealth than other groups. Those differences are what led to social stratification based on four principles.

1. *Social Stratification is a trait of society, not a reflection of individual differences.*

2. *Social Stratification persists over generations.*

3. *Social Stratification is variable. It takes different forms across different societies.*

4. *Social Stratification involves not just inequality but beliefs as well. In the USA, inequality is rooted in the philosophy of the society.*

Robert K. Merton suggested, "Social Stratification[65] in the USA is based upon values of wealth, power, success and prestige" (Merton, 1949). However, everyone does not have an equal opportunity to attain these values. Popularly speaking, being 'cultured' means being well educated and well mannered. This is generally persuaded by upper classes.

It requires credentials such as education, knowledge and communicative skills to attain property, power and prestige. Popular culture is generally left to be pursued by the working classes and the middle classes. This culture normally revolves around TV sitcoms[211], sports, music and films.

British society has experienced significant change since the Second World War, including an expansion of higher education and home-ownership. A shift towards a service-dominated economy, mass immigration, a changing role of women and a more individualised culture has had a considerable impact on the social landscape. In 18th-century Britain: older terms like estates, rank, and order were predominant in society.

In Britain, nobility existed as a distinct social class, but has now integrated itself with those with new wealth derived from commercial and industrial sources. Meanwhile, the complex middle classes have also enjoyed a long period of growth and increased prosperity. The British working class on the other hand, has not been notable in Europe as a magnet for prosperity.

In Britain, the social grade classification created by the 'Readership Survey' over 50 years ago achieved widespread usage in marketing and statistics. It categorises the grades of classification as follows:

GRADE A: Higher managerial or professionals.

GRADE B: Middle managerial or administrative.

GRADE C1: Junior clerical (managerial) supervisors.

GRADE C2: Skilled manual worker.

GRADE D: Semi and unskilled manual workers.

GRADE E: Casual or low-grade workers, Pensioners and others who depend on the state.

In 'Theoretical Framework', the theory of social distinction in the UK was published in 1979. The elite were the gentry by birth and hereditary[181] transmission of occupation, social status and political influence passed down the family. They accounted for 6% of the population and belonged to higher social positions. Those in established and acceptable professions (e.g. academia, law, CEOs and medicine) were considered upper middle class. They constituted 9% of the population. 25% belonged to the middle class. However, 60% of Britons regarded themselves as working class, a phenomenon described as 'working class of the mind'. Traditional British values are still being observed by the elite upper class and upper middle class. Children from the working classes are becoming victims of a 'digital divide', as their parents lack the effectiveness of middle-class parents in helping their children use the internet effectively. Middle-class parents guide and supervise their youngsters, whereas working-class children fall foul of this digital divide. It is in the working classes where traditional mannerisms are on the decline and cultural values are influenced by digital media.

Throughout human history, new technologies for communication have had a significant impact on culture. Inevitably, upon their early introduction, the impact and the effects of such innovations are always poorly understood. The invention of the printing press was at its time perceived to be a threat to European culture, social order and morality. A religious defender of the Spanish Inquisition, Francisco Penna, lamented, "Ever since they began the practice of this perverse[212] excess of printing books, the church has been greatly damaged."

Religion means different things to different people. A simple definition of religion is that it is an attempt by humans to communicate with God, nature, or to reach a state of peace and hope for eternity. Religion affects different cultures in different ways and at different times. When people believe strongly in a given religion, it can have a huge impact on their culture. It can dictate what behaviour and ways of thinking are acceptable in a culture.

Religion and culture are frequently intertwined, but culture cannot be written like a creed[213]. In culture, people are expected to know what to do. For example, in Eastern culture, a guest is treated with a great deal of respect. It is customary that when the guest enters the house, one should stand up, greet the guest and ensure that he takes a seat before the host does. This becomes engrained in a person's actions, such that they no longer need to think about it.

When a society is capable of surviving for thousands of years unless it is attacked from within or outside by hostile forces, often, an attack primarily targets the religion of the society. We have seen this occur throughout history. The most critical point of attack on a culture is its religious experience where one can destroy or undermine religious institutions. Upon this, the entire fabric of society can be quickly subverted[214] or brought to ruins. Many scholars believe that religion is the first sense of community that occurs by mutual experiences, creating a real sense of trust and integrity amongst the people.

The cultural decay in the west today is not haphazard, it was caused. For the last hundred years or so, faith has been beset with relentless attack. In the west, people have been told that it is unscientific, primitive and in short it is a delusion. Beneath delusion lays your own spirituality, your self-respect and peace of mind. Propaganda may have been so successful that you no longer believe that you need peace of mind through the medium of spirituality. Accumulation of wealth, lust and the influence of digital media has deprived man's ultimate wisdom to be spiritually enlightened[82].

When religion is not influential in society, the state inherits the entire burden of public morality, crime and tolerance. Yet governments in developed countries have not been successful in enforcing morality, integrity and self-respect in people. Issues that have risen as a consequence of this include the misuse of freedom of speech, the rise in pornography and general immorality. Crime also continues to increase.

The state has indeed provided benefits through 'social security' but has failed to motivate people to achieve more via spirituality. Instead, the state has chosen to assert dominance through psychiatry and psychology to solve the problems of society. Culture is going through a rapid change in the west. Hence, technology and psychology are used to control and change conditions. However, prisons are filling quickly, and concomitantly[215], churches are losing their former importance in the community.

In the *'modern west'*, our political leaders cannot provide explanations as to why their people are sinking under a tide of vulgarity and sleaze. Anxiety over the collapse of 'respect' can be viewed as a symptom and a cause of a corrosive society. In most urban streets we see a trail of sweet wrappers, sandwich boxes and beverage cans discarded on the grass, littered by youths walking to school. Litter is annoying, but perhaps of greater significance is this modern society that is the abandonment of manners amongst the youth and the death of civility and integrity. We are close to a point where ethical behaviour is regarded as an affliction to be pitied, a loser's burden.

9.2 FAMILY

Family has emerged through sociocultural evolution[118] of kinship[216] groups, from prehistoric to modern times. The family has a universal and basic role in all societies. As a social institution with a biological foundation, the family must be a universal presence. Each society, depending on its demographic structure, economic organisation and religious beliefs will stamp its own characteristics on the family.

Throughout history, cultural interpretations have provided distinctive ways of family involvement in society. Be it the attitudes, culture identity, class, power or ideology. The parenting practices of rituals had effects on the upbringing of the children as well as how the family was constituted. Particularly, a mother's child-rearing[217] style influenced the method of child rearing of the next generation.

In the family history of ancient times child sacrifice, incest and body mutilation were common. Among tribal societies such as INCA, Mesoamerica, Assyrian and Canaanite, religious sacrifices of infants were offered to their gods. The institute of psycho-history has been able to provide data which suggests that in previous centuries, the physical, sexual and emotional abuse of infants and young children proliferated[218] in all cultures (Demause, 1988). Pederasty[219] was both idealised and criticised in Greek literature and philosophy.

It was only in the later Middle Ages that the first child instruction manual and protection laws were brought in, although most mothers still emotionally rejected their children. Until the 16th century, children were treated as erotic objects of adults. It was only in the 16th century, particularly in England, that parents began to refrain from sending children to monasteries[275], nunneries and cheap apprenticeships, in order for their offspring to be controlled. Parents were only prepared to give attention to their child if, in return, they had full control.

The history of childhood is a nightmare from which we have only recently begun to awake. The further back in history one goes, the lower the level of childcare and more likely that children were to be killed, abandoned, beaten, terrorised or sexually abused.

In the beginning of the 18th century, mothers began to enjoy childcare and fathers began to participate in the development of their children. Still, manipulation and beating were used to make children obedient. It is only in the 20th century that some parents adopted the role of helping children reach their own goals in life, rather than imposing upon children the parental wishes.

In China, India, Pakistan, Korea and North Africa, millions of women go missing, and child labour is common. Repression[180] of a child's sexuality is a norm. In those parts of the world women are treated like slaves – a second-class citizen. They are humiliated and emotionally rejected. In India, Pakistan and the Middle East, women go through sex-selective abortions. Societies in the East generally neglect baby girls. Even the grown-up females suffer from passive neglect, such as underfeeding and failure of the guardian to take girls to the doctor when they are sick.

Individual families were normally set up on a patriarchal[191] basis, with the husband and the father determining fundamental conditions and making the key decisions. This ensured the humble obedience owed to the male authority. Patriarchal societies were first developed in Mesopotamian and Sumerian civilizations. Marriages were arranged by parents; the husband served as the authority over his wife and children, just as he did over his slaves. Women were treated as property. It was in those civilizations where they began to emphasise the importance of women's virginity before marriage, and the use of a veil on the face of respectable women when in public. These atrocities were mediated through the tool of religion. Egyptian civilization gave women of upper classes more credit and they witnessed several powerful queens.

Early law was based on the concept of property. Since women were considered property, a man had to be sure who his heirs were, that is, to make sure that he monopolised the sexual activities of his wife or wives. A strong legal emphasis was placed on a woman's sexual infidelity. Since those days, gender inequality has been central to societies for centuries. Women will always be different to men, fundamentally because of their ability to bear children. For some reason, society has decided that there is something inherently inferior about having a female body and producing offspring.

9.3 GENDER EQUALITY IN SOCIETY

According to Bat-Ami Baron, in her book *En*gendering *Origins: Critical Feminist Readings in Plato:* Aristotle portrayed women as morally, intellectually and physically inferior to men. He saw women as the property of men and claimed that women's role in society was to reproduce and serve men in the household. He portrayed male domination of the women as natural and virtuous.

Aristotle had a patriarchal[191] belief system and people have been conditioned to believe that men are superior to women. Greek influence spread with the conquest of Alexander the Great, who was educated by Aristotle. Women's diminished capacity to work during childbearing potentially handed men the responsibility of earning, thus institutionalising the idea of male domination.

Equality between men and women is mentioned in all biblical religious scriptures, but patriarchal[191] enhancement represses[180] this equality. For example, in Judaism, women are endowed with a greater degree of 'binah' (intuition[89], understanding and intelligence). The Ten Commandments[107] require 'the respect of parents, which includes both mother and father'. However, in 'Talmund', some negative things have been said about women. They have been described as lazy, jealous, vain, prone to gossip and capable of witchcraft. All the religious systems imposed patriarchal[191] norms on family life. Jewish women are subject to male pre-dominance under the 'Halakha'. The basic concept of the marriage ceremony is conceptualised[220] as a purchase of the woman by her husband, who takes her as his wife in the unilateral[221] ceremony (Mishna).

In Christianity, it is argued that women should be equal and are made in the image of God. Hence Corinthians 11:3-10: *"I want you to understand that Christ in the head of every man and man in the head of women".*

However, like all religious scriptures, the Bible is littered with complex contradictions. It states in 1 Corinthians 11: *"For if a women does not cover her head, let her also have her hair cut off". For a man ought not to have his head covered because he is "the image of glory of God".*

Corinthians 14:33-35 claims: *"Let the women keep silent in the churches; for they are not permitted to speak... And if they desire to learn anything, let them ask their own husbands at home; for it is improper for a woman to speak in church".*

In 1 Timothy 5:10, the Bible states: *"The women's role is defined in society as, having a reputation for good work; and if she has brought up children, if she has shown hospitality to strangers and if she has washed the saint's feet".* Thus, the Bible clearly has some patriarchal[191] themes across the texts.

In Islam women are not treated as men's equal, although Islam regards women as men's equal. Certain rulers and most legal scholars imposed a system of inequality, which they justified through their interpretation of The Qur'an and Hadith. However, in comparison to other religions, Islam recognises a women's right to choose her own marriage partner. Islamic law regulates a marriage as a contract. It also entitles women to inherit and control their own money.

About a century ago, Western Europe and North America developed a whole set of new values about the way to organise marriage and sexuality. Many of these values are now spreading across the globe. Marriage is supposed to be free of oppression, violence and gender inequality.

Nevertheless, terrible atrocities are committed against untold numbers of women around the world every day and for most of these women, justice will never be served.

Honour killing is a punitive[222] murder, committed by the members of a family or by the wider community. The idea is that a person who has brought dishonour upon the family must be killed as a public statement. A woman is usually targeted for refusing an arranged marriage, being the victim of the sexual assault or seeking a divorce (even from an abusive husband). Bride burning and acid attacks are violent phenomenon that primarily occur in the subcontinent, provoked[252] by disputes over property or insufficient dowries at the time of marriage.

FGM (*female genital mutilation*) procedures, whether for cultural or for religious reasons, are practiced throughout the world, concentrated most heavily in Africa. FGM is performed at any age from infantry to adolescence.

In Europe, since the fall of the Iron Curtain, the Eastern European countries have been identified as a major trafficking source. Young women and girls are lured to wealthier countries upon the promise of money and work, and then reduced to sexual slavery. Three quarters of these women have never worked as prostitutes before.

Across South Asia, untold numbers of infant girls have been murdered by their own families. In modern times, the phenomenon of ultrasound has resulted in gender-selective abortions and has seeded a new method of perpetuating[67] this ancient problem. The liquidation of young babies and female foetuses has created a huge imbalance of gender in the subcontinent and China. A general low status of women in those societies, plus the dowry in marriages and old-age arrangements that favour sons, created this problem.

Baby girls are considered a curse and financial burden, especially to poor rural families, who must cough up expensive dowries upon the marriages of their daughters; while the boys are usually counted upon to take care of their parents in their old age.

In Eastern cultures, even the mothers are brainwashed to comply with these evil practices. Data from the region indicates that one in three women would prefer sons over daughters. Such prejudices lead to aggression and violence against girls as they grow up.

Naeem Tarar, administrator of the Edhi Foundation in Islamabad said, "Discarded babies are found everywhere, some partially buried, thrown in garbage bins and pits, others starve to death."

Tarar also claims that the majority of baby girls are dropped in middle-class neighbourhoods. No matter which class you belong to, in Pakistan every family wants their next child to be a son. Men go as far as marrying multiple wives, just because they believe that a particular woman is 'meant' to only bear girls.

A report in women's news estimated that in India up to five million sex-selective abortions (where the foetus is a female) are performed annually. These are the terrifying pathologies of the educated middle-classes.

Family plays a very important role in society and makes the foundation of the state. A woman is an architect of society. She forms the institution of family life, takes care of the home, brings up the children and tries to make them good citizens. To build a prosperous and healthy society, both men and women ought to have equal rights. Yet, in most Eastern societies and the developing world, women are treated as second-class citizens. They are oppressed in different sectors of life. A woman does not have equal opportunities in areas such as: health, education and gender-biased feeding. She is treated as a commodity, owned by her father and brothers before marriage and then by the husband in subcontinental countries. Even today in some rural and remote areas of Pakistan, customary acts of 'SWARA' are largely prevalent[186]. In these situations, instead of giving blood money as an apology for committing murder of someone from another tribe, an accused family gives their girls in marriage to an aggrieved family as compensation to settle down the blood feud between them. These girls then live the rest of their life as slaves.

The cultural construct of Indian society which reinforces gender bias against men and women, with varying degrees, was identified in favour of male children in a 2011 census. In India, the medical community is providing the illegal service of foetal sex discrimination and sex-selective abortions. India also has the worst statistics for rape, where a woman is raped every 15 minutes.

9.4 FAMILY AND MARRIAGE

The prime purpose of marriage was to bind women to man and thus guarantee that man's children were truly his biological heirs. Through marriage a woman became a man's property.

In the Hebrew scriptures, men were free to take several wives. In ancient Israelite and Arab cultures and religions, several men and prophets were polygamist. Biblical religions allowed a culture of purchasing vulnerable women to turn them into concubines[223], despite these men being married already.

Did God allow the sin of the concubine? Christianity itself demoted the institution of marriage in favour of celibacy[224]. It was only in the 16th century that marriage was reconstituted as an important Christian institute, with the conception and growth of Protestantism. Even then, monogamous marriages were very different from the modern concept of marriage that entails love and fidelity.

Until recent times, love had a little to do with marriage. It was mostly seen as a strategic alliance between families. Men had a wide latitude to engage in extramarital affairs. Any child resulting from this would have no claim to the man's inheritance.

When colonists came to America, at a time when polygamy[190] was still accepted in most parts of the world, the husband's dominance was officially recognised under a legal doctrine[86] called 'covertures[225]'. Under this, the new bride's identity was absorbed into his. The bride gave up of her identity. The rules were so strict that the American woman who married a foreigner, immediately lost her citizenship.

In Eastern cultures and religions, all rules were made up in favour of men in marriages. Women were seen as the weaker sex and were brainwashed through man-made rules. Women were convinced that as long as they were fed, clothed and had a roof over their head, then they must take a secondary place to men. Hindu women treated their husband as 'gods'.

In arranged marriages, it is a common belief that the intimacy will lead to love and happiness. However, the relationship frequently results in adverse consequences. Eastern women, unable to get support from their parental homes, are too nervous to share the truth and too weak to fight for their rights to opt out. They live under a pretence, due to implied familial and social pressures, to accept the relationship and obey the husband in order to gain the acceptance of society.

In 1920, women won the right to vote in the USA. Since then, the institution of marriage has undergone a dramatic transformation. In 1960, state laws forbidding interracial marriages were thrown out, although tradition dictated that the husband still ruled the home. In 1970, the law finally recognised the concept of marital rape, which up to that point was inconceivable, as husbands 'owned their wife'. In Eastern cultures, these progressive steps are occurring at a much slower pace, where the notion of marriage is still to gain dowry, property, wealth and status.

The notion that *'love and marriage go together like horse and carriage'* is a widespread *'ideal'*. Yet the history of marriage shows that this ideal became prevalent[186] only about a hundred years ago. In the history of marriage, it was rare for love to be the main reason to get married.

Marriages in the past were either political unions or merely to produce a biological heir for men. Until very recent times, love had a little to do with marriage, and as such, married men were often involved in extramarital affairs. In many Eastern cultures, love has been seen as a fortuitous outcome of marriage, but not a good reason for getting married.

The importance of love for the maintenance of marriage is greatest in western societies, with higher marriage and divorce rate and with lower fertility rates. In the west, romantic love has made marriages more volatile and uncertain. Accordingly, the number of marriages has been declining. Concomitantly[215], divorces, unmarried partnerships and single-parent families are increasing.

The issue of whether to leave a marriage in which love is not central is a major concern. Marriage is a framework of living that includes other important factors besides love. For example, whether a partner is likely to be a good provider. Long-term happiness and meaningfulness cannot be based upon passive love, but it should include shared activities and profound care. Men may put more emphasis on sex in the relationship, but women emphasise emotional bonding.

Love is a lasting mental and emotional state in which we care deeply for what we love. Love does not come with conditions, stipulations or codes. It is inherently free. It cannot be imprisoned, nor can it be legislated. Love can exist in different forms, be that romantic love, friendship or familial love.

One can argue that for the development of a happy, fruitful and long-term marriage, love, respect and commitment must be shared. These ideas have only recently become the main pillars of a successful marriage. A major moral benefit of this is the equality that is afforded to the women in a marriage, no longer seen as property of the husband to be treated with contempt. This positive evolution[118] of the marriage is far more advanced in Western cultures, where the taboos surrounding mutual divorce of an unhappy couple are less prevalent[186]. Some eastern societies are beginning to show development in their moral view of marriage, with women gaining more rights. However, with the statistics of rape in India, and the stories of female injustice that shroud these cultures, a lot still must be done via the medium of education.

9.5 DIGITAL MEDIA

Our social lifestyle has changed. The world has become a global village. Today it is possible to interact with people living thousands of miles away across the world. Digital media has made physical distances irrelevant. A child of divorced parents can see, talk and interact with both parents, irrespective of who has been granted the custody.

Digital media has the potential to transform our society into a world where there is an unrestricted flow of information. We now spend more time online each day without realising how it affects our everyday lives. What effects does it have on our behaviour and how do we communicate, interact and engage ourselves with people around us?

Of course, digital media has so many positive impacts and has improved our daily lives, but, it has potentially negative consequences. The key is to use it in ways that are beneficial to society, and at the same time avoid risks that arise from malicious[226] use. The most positive effects of digital media are found in the work sphere. It helps to learn and develop businesses professionally as well as aiding collaboration with colleagues and the building of relationships. It also offers opportunities to access the best talents and opens up job opportunities. However, digital media use must be managed well, especially when it comes to extent of usage, the type of social interaction and the nature of its content.

In Justin Healey's book *Social Impacts of Digital Media*, he states, "the rapid uptake of digital technologies hugely impacts on how we communicate, relate, learn, work and spend our leisure time. Digital media literacy is the ability to access, understand and participate or create content using the media."

Those who may not adapt may fall victim to this digital divide. On the negative side, a range of social impacts include: internet addiction, cyber bullying[227], inappropriate exposure to pornography, privacy risks and cyber-crime.

On the internet, things can be posted that are

completely offensive and inappropriate, which most people would never say face to face, for fear of reaction and consequence. Respect, values and morals are slackening. Standards of morality are declining, particularly in sexual permissiveness. People have begun to spend a significant portion of their lives online with other users, experiencing an unprecedentedly new kind of lifestyle. One can argue that the internet erodes our ability to act in concert with locally defined moral standards. We must accept that 'non-geographically bound' communities, such as those online, have replaced traditional communities. Given the fragmentary[228] nature of online communities, will the internet promote the notion of morality? Will it contribute to the development of universally shared models of morality? I fear these are rhetorical[229] questions with negative answers.

Social scientist Christopher J Ferguson claims that the violence in video games has a causal connection to actual violence. A persistent concern about the use of computers and especially computer games is that it could result in isolation and anti-social behaviour. Yet, social networking services can provide an accessible and powerful toolkit for highlighting and acting on issues that affect young people. It can be used to organise group activities to highlight issues and opinions. Consequently, a wider audience could be made aware of these issues.

One major positive that is associated with the rise of interactive technology, namely the internet, is the increased accountability and responsibility that follows people in power. Now, like never before in human history, everyone has a platform upon which to voice their opinions. Thus hypocrisy, prejudice and corruption are all more easily rooted out by the public when the people that govern them sway from their moral duties. This increased means of communication has benefits in countless areas. A more efficient exchange of ideas can have a markedly positive effect on society. For example, communication across the globe within the field of scientific research is rapidly accelerating the unveiling of new discoveries that stand to benefit society.

Ultimately, as discussed above, the internet can be used for good or bad. Its power and influence are unquestionable, and the burden of responsibility rests on the shoulders of the societies that use it. They must ensure that its positives are harnessed[230] for the betterment of the world. Education may be the best policy for safeguarding against the darker side of the internet, although more authoritarian regimes have tackled this problem using other methods. These have included restricting and heavily monitoring the access that people have to the internet (e.g. China).

In a piercing[231] summary of what has gone wrong, Britain's chief Rabbi, Lord Sacks, concludes, "concepts like duty, obligation, responsibility and emotions like guilt, shame, contrition[232] and remorse have been deleted from our vocabulary. Are we not entitled to self-esteem? The small voice of conscience is rarely heard these days. Conscience has been outsourced, delegated away" (Randall, 2009).

Individual misbehaviour is increasing, and this creates an issue not for the perpetrators, but the state. In place of self-restraint, we have installed an all-embracing culture of grievance. Culprits[233] have learnt to claim a victim status. As the banking crises and MP expenses scandal revealed, there is now barely a distinction between legality and morality. In surfing the internet, users are free to view whatever they want, not bound by their own moral values, but rather the fear of getting caught. If everyone else is gaming the system, only a fool would choose to do otherwise. A decent society is not based on rights, it is based on duty, the duty to show respect.

Similar concerns have also been raised in the aftermath of the ascendancy[234] of electronic media; television in particular has been often represented as a corrosive influence on public life. Another member of this electronic media family, the internet, has had a debilitating effect on so-called 'reading brain'. Reading brain describes the short-term gain of using the internet to quickly fulfil a need for information, with no permanent knowledge attained from this behaviour. Maryanne Wolf, an American cognitive[83] neuroscientist, believes that the risk posed by 'reading brain' are particularly impactful on children. She claims, "the Socrates perspective on the pursuit of information in our culture haunts me every day as I watch my two sons use the internet to finish a homework assignment and then tell me they 'know all about it'" (Furedi, 2015).

Apprehensions about the impact of social media on children's brains readily intermesh with alarming accounts of predatory[68] hackers, paedophiles, identity thieves and viruses. The internet serves as a metaphor[29] through which wider social and cultural anxieties are communicated. The internet and particularly the social media bubble are powerful instruments for the mobilisation of people. The relationship between social media and radicalisation is both an interactive and dynamic one. Social media provides a medium through which pre-existing sentiments can gain greater clarity, expression and meaning. The influence of the internet has been most significant in the way it has transformed the lives of young people. Their digital bedrooms symbolise a childhood mediated by social media, mobile phones and the internet. Such interactions have had major cultural consequences.

The digital bedroom emerged as the outcome of the growing tendency to relocate children's activity from outdoor to indoor. The bedroom culture is frequently shaped by children's desire to create their own space and enjoy a level of independence from adult control. Bedroom culture represents the antithesis[80] of the family-centred television viewing in a common room. A growing proportion of children have computers in bedrooms with online access, often misused on the pretence that the internet is required for homework. In reality, they are highly motivated to create a separate space where they can develop their personality, away from parental control. In the wake of this, children create a world that is distinct from that of their parents. Young people attempt to personalise their interests. This results in isolation from other family members. Media technology is used by children to be customised, personalised and consumed privately out of sight of adults.

The electronic culture influences almost every aspect of our life. It offers new ethical values and creates new areas of philosophical reflections. In recent years, our perception of crime and wrong-doing has changed. Technology has actually changed our values, and it certainly has made it possible to behave immorally. Sexual and violent content on the internet can be viewed without anyone ever knowing. No one needs to know what you are seeing or who you are corresponding with. Adultery sites are common as well as pornography sites and other violent photos. You can slander[235] anyone, with no proof and no consequences.

According to a study in the UK by the Institute for Public Policy Research, Britain's teenagers are among the most badly behaved in Europe. The Institute explains that Britain's dismal record is due to a collapse of family and community life, pointing out attitudes of no respect, no morals and no trust. Changes in family structure have had an impact on our care for the elderly. According to Age UK, starting families later and having fewer children may have an effect on the availability of parental care (Age UK, 2017). Greater geographical separation of families and increased divorce rates have increased the numbers of people living alone. Reduced family size reduces the number of children available to be the carers. In the UK, family attitudes are changing fast, in comparison with the rest of Europe. In Italy, 93% of 15-year-olds regularly eat with their parents without a digital device around.

In the UK, it is almost crazy to believe that we are following America's code of conduct amidst their own moral collapse. This moral collapse in America is evident through a report by the Centre for Disease Control and Prevention: one third of the entire population currently has a sexual transmitted disease. It costs America 16 billion dollars a year to treat STDs. Surely, the government must be receiving more in taxations from companies involved in pornography. However, the American media is so good in their propaganda that around the world they are considered forward going.

Similarly, the British state rewards unmarried mothers with a level of benefits that most would be unable to earn in legal employment. They are incentivised to go solo via no sense of right or wrong, such that this incredibly personal decision becomes political rather than emotional. Having more children with a variety of fathers means a rising tide of benefits and handouts. According to Lord Sack, "without a shared moral code, there can be no free society. Either we find the moral sense, or we will find that in the name of liberty, we have lost over freedom."

Vikram Dodd of *The Guardian* reports that leading experts have dismissed a claim that Britain has one of the highest crime rates in the developed world. Barry Irving of the police foundation called the report 'simplistic', claiming that it failed to take into account changes in the reporting and recording of crime. However, one could argue that the police are using pedantic[236] and tedious arguments to deflect the clear problems relating to increased crime rates. This same report argues that a moral decline resulting from the 60s cultural revolution has fuelled criminality. It continues, suggesting that the real problem is the loss of international moral principles that prevent people from committing crimes in the first place: "Young people who grew up in troubled and dysfunctional households in which moral values are not implanted, who attend schools where teachers are afraid or unwilling to teach the differences between right or wrong. These young people live in communities in which the influence of religious faith is negligible, therefore, they will naturally be drawn towards the self-gratification."

Our liberal establishment has been ignoring links between family breakdown and social disorder. Family breakdown associated with divorce or separation is a major cause for poor mental health. However, a governmental mental health policy launched recently makes no mention of the effects on children's mental health of conflict between parents and living in fractured families. Conflict between parents has been associated with an array[237] of adjustment problems in children. This results in concerns relating to conduct, depression, anxiety, low self-esteem and behavioural difficulties.

CHAPTER 10

THE FORMATION OF STATE

10.1 EARLY THEORIES

There are numerous different theories and hypotheses[238] regarding early state formation, ultimately attempting to explain why the state developed in some places but not others. One theory is that different groups of people came together to form states because of shared rational interest. The theory largely focuses on the development of agriculture, and population and organisational pressures that followed. An answer to these issues may have been seen as state formation.

One of the most prominent theories of the early formation of a state is the 'Hydraulic Hypothesis', which asserts that the state was the result of the need to build and maintain large-scale irrigation projects. The theory is that in arid[239] environment, farmers would be equally restricted by the production limits of small-scale irrigation. Eventually, different agricultural producers would join in response to population pressure and the arid environment. They would create a state that could build and maintain large-scale irrigation projects, with surplus food stocks. This was a direct consequence of agriculture and the division of labour. Another hypothesis of early state development is that long-distance trade networks created an impetus for the state to develop at key locations: such as ports or oases.

Another, perhaps more conflicting theory of state formation emphasises that dominance of some populations over other populations was the key to the formation of state. Subscribers to this viewpoint believe that the state formed due to some form of oppression by one group over others. In general, this theory highlights economic stratifications[65], conquest of other people and conflict in circumscribed[240] areas were the basis of the formation of a state.

Friedrich Engels argued that "*The origin of the family, private properly and the state developed because of the need to protect private property*" (Engels, 1884). The theory claimed that surplus production of agriculture created a division and specialisation of labour, leading to different classes. These classes included those who worked the land and those who could devote time to other tasks. The need to secure the private property of those living on surplus production produced by agriculturalists resulted in the creation of the state. Social stratification was the primary reason for the creation of the state.

'Neo-evolutionary theory' argues that *"a human society evolved from tribes and chiefdom into states through a gradual process"* (*Encyclopaedia Britannica*, 2018). Groups that gained power in tribal society gradually worked towards building the hierarchy[122] and enforcing segmentation[241] for creating the state. Bureaucracy evolved to support the leadership structure in tribes and used religious and economic stratification[65] as a means of consolidating power. War may have played a key role in this situation, because it allowed leaders to distribute benefits in ways that served their interests.

Technological, religious and social development is crucial for state development. However, most of these factors are found to be secondary to the need for defence from military conquest and a level of military organisation for conquering other states.

Through history, political entities created by humans have expanded from basic systems of self-governance into a monarchy and then to the complex democratic and totalitarian regimes that exist today. In parallel, political systems have expanded from vaguely defined frontier-type boundaries to those visible today. Early dynastic Sumer and early dynastic Egypt were the first civilizations to have defined borders, in roughly 3000 BCE. By 2500 BCE the Indus Valley civilization located in modern-day India, Pakistan and Afghanistan had formed boundaries extending

600km inland from the Arabian Sea. 336 BCE saw the rise of Alexander the Great, who forged an empire from Greece to the Indian subcontinent, bringing Mediterranean nations into contact with those of central Asia as the Persian Empire had before him.

The European states of the Dark Ages and Middle Ages gained their authority from the Roman Empire, and modern democracies are based in part on the example of ancient Athens. In Middle-Age China, the Tang dynasty took control of China and extended its border from Eastern China to the north. The borders were redefined many a time because of the attack from neighbouring Tatar (Mongol) tribes.

After the death of the Prophet Mohammad in 632 CE, the Qur'an and the teachings of Islam inspired the genesis[242] of a new civilization. In less than a century, the Islamic Caliphate rapidly extended its reach from the Atlantic Ocean and Andalusia in the West to Central Asia in the East. The subsequent Muslim Empires of the Umayyads, Abbasids, Fatimids, Ghaznavids, Safavids, Mughal and Ottomans were amongst the most influential and distinguished powers in the world during Middle Ages.

In Western Europe around 800 CE, England, Iceland, Norway, Germany and Italy were all recognised as nation states. Until then they existed as a sole entity under King Charlemagne. These states emerged through ethnic, linguistic and geographical differences. In 1299 CE, the Aztec and Maya Empires gained prominence in lower Mexico and lasted over 300 years. In the late 13th century, central Asia witnessed the rise of the Mongol Empire, which extended across Europe and Asia, reigning for 370 years. Following its dissolution[243], the Ottomans exploited the opportunities that the Mongols left behind.

After the 'Classical Period' (the Middle Ages from 500-1500 CE), a period that oversaw the rise of Christianity and the Islamic Golden Age, the *'Age of Enlightenment and Discoveries'* followed. The mid-15th century invention of modern printing revolutionised communications, as civilisation entered a scientific revolution with the accumulation of knowledge and technology. Just before the modern period of the industrial revolution, monarchical systems began to collapse with the start of the Magna Carta in Britain and the Russian revolution. Democracy began to gain momentum in the west. Communism, capitalism, socialism and democracy emerged, ending the traditional kingdoms and Imperial powers.

In 1770, the United Kingdom's military opened fire on the North Americans. This event became known as the Boston Massacre. Having lost this battle, 1783 saw the British sign the 'Treaty of Paris', recognising American independence. The borders of this fledgling[244] country expanded over time from east to west. History shows that nations often start out as city states, expanding outwards under the leadership of an ambitious ruler. These rulers plough forward gaining all the land that surrounds them and are only stopped by huge natural barriers such as oceans, mountains and deserts. These are the very geographical features that often form natural borders between different nation states. For example, the Rio Grande River splits Mexico from the United States of America, and the Rhine River splits Germany and France in Europe.

10.2 STATE SYSTEMS

There are three main political systems that various states of the world have adopted:

- **'Democracy'** is a political system of governance either carried out by the people directly or by elected representatives. In democracy, free enterprising is allowed, which means that people or groups can own their own business. This system leads to rich and poor in society. Democracy is based on the principles of equality of rights and freedom of speech and choice.

- **'Communism'** is an ideology of economic equality through the eradication[74] of private property. They believe that inequality and suffering results from capitalism. It is a socioeconomic structure that stands for the establishment of a classless and egalitarian[189] society.

- **'Capitalism'** is an economic system based on private ownership by a means of production of profit. Characteristics central to capitalism include private property, capital accumulation, waged labour and pricing systems for competitive markets. The means of the production and distribution of goods are owned by a small number of people, known as the 'Capital Class'.

Private ownership is not allowed in a communist system, instead power is invested in a group of people who decide the course of action. The community solely owns the resources or the means of production. Whilst the profit of any enterprise is equally shared by all the people in a communist society, the profit in a capitalist structure belongs to the private owners. It is the private parties who control the resources. Supply and demand, as well as price, are determined by the quantity of the production and how many people want them. Prices are increased or decreased according to demand.

In communism, prices and their levels are fixed according to the need of the society. Carl Marx, the father of communism, once stated that "religion is the opium of the people". He rejected religion because he saw it as harmful, preventing people from seeing class structure and oppression around them. Thus, religion was used as a political tool for the suppression of this communist revolution. Carl Marx believed: "Religion does not make man, it is man who makes religion." He believed it was a lack of self-confidence and self-esteem that caused people to choose religion. Therefore, some communist countries that followed the Marxist-Leninist dogma[245] were atheistic and anti-religious. Nevertheless, religious communist groups also existed in communist countries. In a communist system, by law one can practice religion but not preach it. Capitalism is the only system that allowed freedom of religion, including freedom from it. Thus, capitalism neither supports, nor opposes religion, provided those religious practices do not violate the rights of others.

A dictatorship is a form of government where a country or state is ruled by one person or political entity. It is exercised through various mechanisms to ensure that the entity's power remains strong. A dictatorship is a type of authoritarianism in which politicians regulate nearly every aspect of public and private behaviour of their citizens. They generally employ political propaganda to decrease the influence of alternative government systems. In the past, different religious tactics were used by dictators to maintain their rule. Perhaps the archetypal example is monarchy systems of the West. In the 19th and 20th centuries, traditional monarchies gradually declined and constitutional democracy emerged. Inheriting power through family ties, military dictatorship or self-coups are all dictatorships as they suppress democracy to attain office.

The quality of the government, regardless of which system of governess it chooses, is judged on the systems which offer its citizens the opportunity to live securely. It should provide a dutiful and obligatory binding through provision of a civil service, local government, police, courts, education and health service.

For the earnings of the state, the relevant major revenue sources are income tax and sales taxes. The government legally require their citizens to hand tax through proper channels. States also borrow money as debt and issue bonds to those who lend them the money. The government also print money and put it into their accounts, either directly or indirectly. None of these methods of raising money is immoral or irresponsible. They also collect profits from state-run enterprises and seize funds from other countries that they conquer.

Although the notion of peace and prosperity are hopelessly complicated, after the Second Word War, the United Nations was founded with its prime objective to secure and safeguard the sovereignty of all nations from unilateral[221] attacks by powerful countries. It has failed miserably in this endeavour, and has not lived up to its responsibilities, largely due to the veto[246] system used by the six permanent members of the security council. In general, the USA and Russia, the most powerful countries, impose their right to a veto on major policies, thus accepting violation[247] of world peace through their vote. A strong propaganda machine is operated via digital media on the one hand, and the same countries supply terrorist organisations with arms on the other hand.

10.3 BANKING

Let us assume state rule over a territory that has developed beyond the stage of primitive barter[248] economy, where money is used as a common medium of exchange. As a state ruler, you can in principle, confiscate whatever you want and provide yourself with unlimited funds. Emperors, kings and government heads have been doing this for thousands of years, collecting various taxes that they levied[249] in the common interest of providing security and social facilities. To avoid encountering difficulties, the state has promoted the banking system and has freed itself of the constraints[250] of lending money or to collect taxes. At the same time the government has taken control, such that only they can produce money. Thus, the state has the monopoly to produce worthless pieces of paper to replace quantities of gold and silver. The beauty of the system is that marginal cost of producing the paper money is almost zero, therefore, the purchasing power of this money is zero as well. Thus, money through banks has become a magic wand for institutional set-up for governments.

Banking is defined as an organisation that provides a facility for acceptance of deposit and provision of loans. In the ancient world, banking began as grain loans to farmers and traders who carried goods between cities. It began around 2000 BC in Assyria and Babylon. It would later develop in Greece. During the Roman Empire lenders, based in temples, made loans and exchanged money.

Financial history in ancient China and India also shows evidence of money lending after this period.

The development of the banking system during the Medieval period and Renaissance[251] Italy, in cities such a Florence, Venice and Genoa, was established by the Bardi and Peruzzi families. In the 15th and 16th century, banks were established in Northern Europe. During the 20th century, rapid developments in telecommunications and computing caused banks to dramatically increase in size and operations. The financial crises of 2008 caused much debate and provoked[252] bank regulations.

Historically, early religious systems in the East did not forbid interest of any kind (usury[253]). Hence in ancient times, if one lent 'food money' or monetary tokens of any kind, one did so to charge interest. In the Torah, interest taking was criticised, but interpretations of the Bible vary on this matter greatly. One common understanding is that Jews are forbidden from charging interest on loans made to other Jews but obliged to charge interest on transactions with gentiles[254]. As stated in Deuteronomy[255] 23:19 of the Old Testament: "Thou shalt not lend upon interest to thy brother: Interest of money, victuals[256], interest on any other." Following this, Deuteronomy 23.20 states: "unto a foreigner thou must lend upon interest."

However, laws against usury[253] were condemned by many prophets. It was the interpretation that interest could be charged to non-Israelites that would be used in the 14th century by Jews living within Christian societies. Originally, the charging of interest was banned by Christian churches. However, over time this charging of interest became acceptable. The rise of Protestantism in the 16th century weakened Rome's influence, freeing up the development of banking in Northern Europe.

In Islam it is strictly prohibited to take interest; 3:130 of the Qur'an states "and Allah has permitted trade and has forbidden interest." Riba[257] (a Jewish term for charging interest of any kind) is also forbidden in Islam. Riba is either of the following two types:

1. An increase in capital with no service provided.

2. Commodity exchange in unequal quantities.

Trade in the form of promissory[258] notes is prohibited in the Qur'an. Islamic banks do not charge interest, but this could be done through charging of loans in different ways, with different ownership models of risk sharing.

At the start of the 15th century, the Medicis were Europe's greatest bankers. Political power distracted them from the highly focused business of money makers. The Medici family later triumphed, becoming dukes of Florence. Their role as leading bankers was surpassed by the Fugger family of Germany. The Fugger family have amassed great wealth by managing the finances of the Pope and the great princess. The shift of European power to Habsburg is the basis of the Fugger family wealth. In 1487, they made their first loan, taking as security an **interest** in silver and copper mines in Tyrol. In 1491 a loan was made to the Roman Empire at the time, known as Maximilian. In return, Habsburg secured the feudal rights to two Austrian countries. Maximillian's grandson, Charles, wanted to succeed his grandfather as German king and the Holy Roman Emperor. The post involved elections and there was a rival candidate, the French King Francis.

10.4 BANKS AND STATES

Charles turned to the Fugger family for his election expenses. This required 852,000 Florins, to be spent on bribing the 'seven electors'. Of this massive sum of money, the Fugger family provided two thirds, and Charles was duly elected. Under normal circumstances, interest rates of the time never exceeded 12%, but the urgency of the loan meant that the rate could rise to 45%. Charles was not able to foot such a large debt to the Fuggers, and so the Fugger family gained a Spanish order of knighthood, together with profits from the mercury and silver mines.

The 17th century saw the rise of the Rothchild banking dynasty. They saw an opportunity in that until now, banking was only used by the upper class, namely kings and emperors. The Rothchild family were the first to provide their banking service to the citizens, introducing a simplified version of a bill of exchange. The family made another key observation; the money left on deposit by a customer is a large sum, and negligible to that which is usually required for withdrawals. They could use the collective customer deposited money to provide loans for clients; these clients would then pay the bank with interest. Thus, the Rothchild bank could make profits on all their loans.

In 1656, Johan Palm establishes the Stockholm banco. In consultation with the government they issued credit notes, which could be exchanged on presentation to the bank, for a stated number of silver coins. They also started bank notes as a genuine currency.

In 1694, a joint stock company turned into 'The Bank of England' by arranging a loan of £1,200,000 to the government. During the 18th century, they became a central bank by organising government bonds[259] when funds needed to be raised. For the purpose of being a national bank, they needed a large reserve of gold, which they accumulated until almost the entire hoard of the nation's bullion was stored in its vaults.

In England in the 18th century, Goldsmiths accepted money as deposits from customers purely for safe keeping. However, they began to lend some of this money out to clients and this business grew until, as the 18th century reached a close, the Goldsmiths made banking their sole business.

10.5 HOW THE MONETARY SYSTEM WORKS

Money can be used to purchase goods at the marketplace, thus, in the past the money was backed by precious metals, like gold or silver. Today all western currencies are backed by debt. Money is simply a 'promise return' on a piece of paper. Money can only be 'created' when somebody signs a debt obligation. There are two scenarios:

1. If a company issues a loan, a person can borrow money from private individuals. In this case no new money is created, the money just changes hands.

2. Money is created by the Central Banks. For example, the Federal Reserve Bank of the USA or European Central Bank. The Central Bank does not need to have the money to pay the government. It must just buy a 'bond[259]' of the government. Since the Central Bank have the monopoly, they can effectively create money out of thin air. The 'bond' is seen as collateral and the government receives the money from the Central Bank.

A way of defining a bond is as follows: If a company issues a bond, the money they receive in return is a loan, and must be repaid over time. So, the Central Bank buys a bond from the government, simply printing out money and loaning it. The government must then repay this money with interest. In the 2007/2008 banking crisis, 'subprime loans' were bundled over to the 'FED' in exchange for money created from thin air.

If someone takes a loan, it requires a payment of interest together with the loan. Thus, in a debt-based monetary system, the amount of money in circulation is forced to grow indefinitely due to the interest mechanism. So, the government must constantly increase the amount of 'national debt' in order to inject new money into circulation for the debt to be cleared. This can be described as a vicious cycle, or a paradox[91] that will never come to completion. As the government takes a loan to try to clear its national debt, interest on this loan increases the debt. The bottom line is that the national debt can never and will never be cleared.

In the case of smaller countries, issuing new bonds[259] becomes increasingly difficult if the potential lenders lose faith in the ability of the government to pay this money back. A thought experiment can be performed here for one to understand: A country has a banking system that lends money with interest to the citizens, thus as time goes on the bank makes money, taking it from circulation before lending again. However, there comes a point when the back lends money with interest, but there is no longer enough money in circulation for this to be returned. In order to prevent this situation, the government borrows money from the bank and injects it into the economy. However, if this government cannot acquire a loan from the bank, as is often the case for less wealthy countries, less money is in circulation as the bank has soaked up the money. Consequently, the value of everything increases in a process known as inflation[260]. In other words, money rapidly loses its purchasing power.

10.6 MODERN WAYS OF BANKING

A national bankruptcy will end up in 'Monetary Reform' resulting in governmental bonds being devalued. In the liquidity trap of 2008-2012, the Bank of England pursued quantitative easing (printed money to loan and rescue banks) but this only had a minimal impact on underlying inflation[260]. This is because although the bank saw an increase in their reserves, they were reluctant to increase bank lending.

The Bank of England is the custodian to the official gold reserves of the UK and around 30 other countries. 'The Vault', beneath the City of London, covers a floor space greater than that of the third-tallest building of the city, Tower 42, and needs keys that are three feet long to open. As of April 2016, the bank held around 400,000 bars, which were estimated in January 2017 to have a current market value of £161,000,000,000.

In April 2009, the 'International Monetary Fund' (IMF) held 3,217 tons of gold, which has been constant for several years. The IMF is an international organisation with its headquarters in Washington DC. It came into existence in 1945 with 29 members countries with the goal of reconstructing the international payment system. It now plays a central role in the management and balance of payments, difficulties and international crises. Countries which face balance or payment problems can borrow money.

The IMF has three primary functions.

1. Fix the exchange rates.

2. Provide short-term capital aid for balance of payments.

3. Provide economic growth and projects. for infrastructure. The IMF has created a policy such that it has a monopoly on the natural resources of the country that it lends money to. If the country fails to meet its quarterly payment, it must comply with these demands of the IMF (payment is in the form of the country's natural resources).

Some of these conditions are austerity[177], resource extraction, devaluation, price control and governance.

10.7 CORRUPTION IN DEVELOPING COUNTRIES

Historically, the managing director of the International Monetary Fund (who is the chairman of the executive board), has always been from the United States or Europe. Amongst the 24 executive directors, only eight countries can appoint these positions. These countries include: the US, Japan, China, Germany, France, the UK and Saudi Arabia. The remaining directors (16) represent constituencies consisting of 4 to 22 countries.

Since the 1990s, there has been an explosion in the exposure of corruption amongst the leaders of the developing world. The IMF has actively endorsed the use of capital control under some circumstances. In 2011, the IMF went so far as to recommend guidelines and governance structures regarding when nations should deploy capital controls.

In the developing world, corruption is a complex phenomenon. Its roots lie deep in bureaucratic and political institutions. There are numerous cases of abuse of public office for private gain when an official accepts, solicits[261], or extorts[262] a bribe. Public office is also abused for personal benefit even if no bribery occurs through patronage, nepotism, theft of state assets or diversion of state revenues. Bribery in the private sector should be a concern of the bank. This is because the bank lends to the government and supports policies, programs and projects. The IMF is promoting political and bureaucratic corruption by lending money to these countries. In return it acquires political control over these states.

Why does the IMF lend money to these governments, where institutions are weak and government policies do not favour auditing and accounting? It is the IMF's duty to base its approach of granting loans to these governments on evidence and analysis of corruption in these developing countries.

10.8 POVERTY IN AFRICA

Although the African continent is blessed with gold, oil, diamonds, coltan, uranium and other valuable resources, its inhabitants have long been amongst the poorest. So many international organisations and foreign governments have been trying to promote regional development, food production, education, better housing and healthcare. Yet, the African governments often appear clueless when it comes to lifting their people from extreme poverty to enjoy economic growth.

Everyone seems to have a favoured explanation for this tragic phenomenon, citing pervasive[263] corruption, dysfunctional democratic institutions and inadequate justice systems. Multinational corruption, ineffective international aid agencies and warlords each must bear responsibility for this sorry situation. In 2010, fuel and mineral export from Africa was worth $333 billion, more than seven times that of the aid received by Africa that year. This is before you factor in the vast sums involved in corruption and tax fiddling.

The empires of colonial Europe and the cold wars have given way to new a form of dominion. Tom Burgis, in his book *The Looting Machine*, quotes: "New empires controlled not by nations but by the alliances of unaccountable African rulers governing through shadow states, Middlemen who connect them to the global resource economy. Multinational companies from the West and the East that clock their corruption secrecy, prefer not to think of the mothers of Eastern Congo, the slum dwellers of Luanda and miners of Mavange. As long as we go on choosing to avert our gaze, the looting machine will endure" (Burgis, 2015).

Africa is not poor, it is just poorly managed with a slavery mindset rooted in its complex of inferiority. Colonialism cultured this vulnerable attitude, of the African people, that has persisted beyond independence of these nations. The rest of the world has exploited this for their own personal gain ever since.

10.9 WHY POOR COUNTRIES STAY POOR

The UN Human Development Department reports that 1% of the richest adults account for 40% of global assets. Oxfam International released a report in 2013, stating that the richest 1% own 40% of global wealth.

Two groups of countries known as 'The Twin Peaks' (Global Gini Coefficient) account for most of the world's wealth. The first group has 13% of the world's Purchasing Power Parity (PPP) income. The group includes the USA, Japan, Germany, the UK, France, Canada and Australia. The second group has 42% of the world's population and receives only 9% of the world's PPP.

The middle third of the distribution is the domain of China and the bottom third of global wealth is the domain of India. The richest 1% owns 58% of total wealth in India. An Oxfam report in January 2017 showed that 57 billionaires in India now have the same wealth as that of the bottom 70% of the country. In neighbouring Pakistan, according to a UN Development Program report, economic inequality is rising in Pakistan. The country's institutions continue to benefit the rich and burden the poor. The report alleges that Pakistan's response to inequality has been superficial and focuses on symptoms rather than root causes.

Increased inequalities and corruption on a large

scale creates poverty and extreme riches. The curse associated with being a nation of plentiful natural resources suggests that a resource windfall generates additional wealth, which raises the prices of non-tradeable goods, but the manufacturing sector shrinks.

Countries get rich by finding a niche[264] where they have a comparative advantage, specialising and then producing whatever their speciality is for the world markets. They diversify trades within the cities, which creates synergy[265]. The government then supports growing businesses with monopoly rights and tariffs to prevent cheaper imports. Out of these hot-spots of business activity, innovation, manufacturing and subsequent exports start to blossom. No country has ever developed without manufacturing. These countries in the developed world have constitutions and laws to provide their citizens with social welfare, which includes personal security, free healthcare and free education for all. They make it possible to impose adequate taxes on earnings at various levels of income. The government promotes free enterprise and encourages banks to support innovative projects through affordable financial packages.

On the other hand, they encourage underdeveloped countries, rich in natural resources, to obtain funds through the International Monetary Fund (IMF). The IMF play their roles by encouraging developing countries to gear their entire economy towards their raw materials. They provide loans to developing countries and encourage them to focus on tradeable commodities.

Those countries are forbidden from using tariffs, subsidies or other forms of market protection to develop any kind of manufacturing. This is a false route for economic progress. Poor countries are poor because they are persuaded to grow raw materials and their leaders are tempted with promises of free trade and loans. In the past 30-40 years China and India have made progress by completely ignoring IMF policies.

The IMF can help countries, whether rich, middle-income or poor, to provide loans at affordable terms. It appears that in the case of poor and underdeveloped countries, the IMF lends money and gains control of the broad economic policies of the country, along with imposing methods of debt restructuring.

Countries that approach the IMF are those whose international reserves are depleted, their economic activities are stagnant or failing. IMF loans are typically disbursed[266] into several instalments, with each instalment conditioned that agreed targets be met.

Systematic corruption in poor countries can promote a source of conflict and a threat to economic growth. The 'OECD' (Organisation for Economic Cooperation and Development) suggests that the best way of achieving greater equality is to focus on raising the living standards of the poorest 40%. Their governments ought to invest in skills and education as well as promoting better quality jobs.

Even in developed countries, the 'rules of the game' are stacked in favour of monopolists. These

CEOs of corporations get salaries and bonuses that are roughly 300 times that of ordinary workers in the same corporations. No increase in productivity justifies this sort of difference. The gap between rich and poor, together with the growing income inequalities is the biggest risk that the world may face in developed and emerging economies.

CHAPTER 11

EDUCATION

As societies developed, for hundreds of thousands of years, children educated themselves through self-directed play and exploration. The strong drive in children to play to develop, came about during our evolution[118] as hunter-gatherers, who allowed children to develop naturally and let them learn the skills of tool making and hunting gradually.

As agricultural societies grew, the people started to live in dwellings, where crops were planted as opposed to living a nomadic life. Successful farming required long hours of relatively unskilled repetitive[267] labour, much of which would be done by children. Children became forced labourers and consequently play was suppressed. Wilfulness, which had been a virtue, became a 'vice' that had to be beaten out of children.

Education is the process of facilitating learning

and pursuing the acquisition of knowledge, skills and values. Initially the idea of being educated came from ancient Greece in around the 5th century BC. The word 'school' comes from the Greek word 'schole' which means leisure. In those days leisure was synonymous[268] with learning. Those schools were elementary for teaching arithmetic and reading. From there came Plato's Academy and Aristotle's Lyceum[269]. Plato's philosophical thoughts about education are mentioned in 'The Republic: 536 d-e': he states, "a free soul ought not to pursue any study slavistily," Plato highlights an issue here that remains recognisable to this day. His emphasis is on leisure rather than hard study.

After the ancient Greeks, Roman schooling broadly followed the Greek model. Schools were taken up by Arabs. The cultures that have achieved greatness are those where learning was fostered[270] for the sake of gaining knowledge.

The Islamic golden age dates from the 8th to the 13th century. Science, economics, agriculture, philosophy and medicine scholars from various parts of the world gathered and translated the world's classical knowledge into the Arabic language. During that period education was considered a central pillar. Primary schools were attached to mosques known as 'Madrassas'. These institutions multiplied throughout the Islamic world. Educational attainment as qualifications were granted by the scholars rather than the institute.

The University of Al Karaouine, founded in 859 AD in Fez Morocco, is the world's oldest. They awarded attainments in medicine, ophthalmology, engineering and social sciences. These 'attainments' were equivalent to today's degrees.

Looking back, building the pyramids[271] must have required a good grasp of mathematics and geometry. The builders of pyramids held no degrees. In an article published in Harvard magazine, 'Who Built Pyramids', it has been claimed that they were testament to the engineers and to decades-long labour. The Sphinx[272] is carved directly from Giza and sits below the surface of the surrounding plateau[273].

The invention of the printing press in the 15th century made a significant impact on literacy. With printing came the development of thought and further interest in literature and science. Out of that emerged the 'Renaissance[251] era'. Learning became firmly subject centred rather than child centred. In Europe, Latin was the language of learning and most schools were devoted to it; voices of concern started to emerge, saying that things could be different if the children were taught without compulsion[277] and fear. A Czech teacher, Comenius (1592-1670), argued as a 'champion the universal education' for the 'commonality of the education'.

In *Education: A Very Short Introduction*, Gary Thomas writes: '*Today education acts as a socializing agency*, with schools and educational institutions judging children and adults according to the same universal standards. These standards are applied equally to all people irrespective of the unique character of the individuals. Compulsion and fear have been replaced with degrees which may provide a better job opportunity.'

Children's lives changed from free pursuit of their own interest to slave labour. Agriculture and the associated ownership of land created clear status differences. People who did not own land became dependent on those who did. That 'status-conscious' society gave birth to kings, chiefs and feudal lords.

In the Middle Ages, lords and masters had no qualms[278] about physically beating children into submission. In western societies, with the rise of industry, a new bourgeoisie[279] class developed. Although feudalism gradually subsided, exploitation of young children carried on. Children were working all hours in beastly conditions. Unfortunately, this kind of exploitation is still being practiced in many parts of the world, particularly where there is a 'feudal' society in place. A good child was an obedient child, one that suppressed his or her wishes, opting to carry out the orders of the adult masters.

The introduction of education in Europe through Moorish Spain roughly 1,000 years ago triggered an intellectual odyssey[274] and transformed the concept of education.

Throughout the Middle Ages, the Muslim world was more advanced and more civilized than Christian Western Europe. As most Muslim schools were attached to Mosques, in Europe the foundation of education was through Christian monks. It was through these monasteries[275] that universities evolved. Initially the universities in the Islamic and Christian world were built by religious leaders and organisations to promote learning and to mediate[276] the spread of faith.

In the view of some scholars, 'reformation[280]' was a driving force. Much of the impetus of universal education came from the emerging 'Protestant' religion. For various reasons, some religious and some secular, the idea of 'universal compulsory education' arose and gradually spread. As industry progressed and became somewhat automated[281], the need for child labour declined.

In Europe, the idea began to spread that childhood should be a time for learning, and schools emerged as a place for learning. By the 11th century, aristocratic families would send their children to established schools outside of religious organisations. These schools taught Latin, English and French. Poor children were sent to become apprentices in various skills. The most crucial concepts that schools promoted were punctuation, obedience, tolerance and long hours of tedious work.

11.1 EDUCATION IN THE THIRD WORLD

At the time of American Independence in 1776, less than half of white children and almost no black children went to school. Today almost every

American child finishes primary school, and all go to the secondary schools. The USA didn't achieve universal primary education until 1900. At the time the average American earned \$4,500 a year, in today's dollars, which is six times as rich as the average person in Indonesia, and fifteen times richer than in Uganda.

Even today many countries lag behind where the US was in 1776; in Ethiopia 31% attend primary schools, in Sudan 53% and in Niger 30%. More than 115 million children are not in schools in the developing world. They start school but do not complete primary education because of poverty.

Positively, in some of the poorest countries, such as Burkina Faso, Madagascar and Nicaragua, school attendance rates are rising faster than they ever have. Countries such as Indonesia, the Philippines, South Africa, Peru and Jamaica are on the cusp[282] of reaching universal primary education. In the developing world, mass schooling started in 1948, through the Declaration of Human Rights. So there has been a gap of 100 years between the developed and underdeveloped world.

The brute force methods long used in factories and on farms were transported into schools to make children learn. Some of the underpaid, ill-prepared school masters were clearly sadistic[283]. Everyone assumed that to make children learn, their wilfulness would have to be beaten out of them. Punishments of all sorts were understood as intrinsic[284] to the educational process. In recent times, methods of schooling have become less harsh, but basic assumptions have not changed. Learning continues to be defined as children's work and power assertive means are used to make children do that work. In the 20th and 21st centuries, methods of discipline became more humane, or at least less corporal. Lessons have become more secular as the curriculum is expanded and knowledge expands.

Today, not all societies channel sufficient resources into support for educational activities and institutions. However, it has become apparent that they at least acknowledge the importance of this. It is obvious that children are born illiterate, innumerate and ignorant of the norms and cultural achievements of the communities or society. With the help of professional teachers and dedicated family, in a few years they can read, write, calculate and act in culturally appropriate ways.

Some learn their skills with better facilities than others, and so education also serves as a sorting mechanism for social classes. This obviously has an enormous impact on the future economic fate of the individual.

Behind every school and every teacher is a set of related beliefs that influences what and how the students are taught. This philosophy answers the question about the purpose of schooling, the role of teachers, what should be taught and by what method. A teacher-centred philosophy tends to be more authoritarian and conservative with emphasis on the values and knowledge. This gives power to the teacher to choose the curriculum, organise the school day and construct classroom activities.

11.2 THE ACADEMIC DEGREE

To obtain a degree has become a central part of the American dream – "Go to college, get a job, buy a house, and raise family." Getting a degree is all about opening opportunities. I believe that a degree is a form of recognition and prepares you, both intellectually and socially. It also supposedly enriches your life, by increasing your earnings.

Of course, there will always be people who are motivated by the thirst of accumulating knowledge through natural curiosity and the love of individual pursuits. Those people either become a PhD student or join an organisation where research is appreciated and rewarded.

However, lacking an educational degree does not mean that a person will not succeed. A degree is an achievement and not indictive[285] of a person's character or ambitions. Many people in our history have succeeded, some far beyond general standards, without having a degree. A university degree in education gives people a rounded experience of academic study.

To become a doctor or an engineer you need to understand all the procedural knowledge, transferred over centuries. In addition to that, one requires practical exposure to the problems, under supervision while you implement that knowledge.

Both engineers and doctors provide important services. Educational verification is necessary for the above two professions in the form of official documents.

I have been trying to enter debate about the importance of the university degree. This is because, in the past 30 years, I have seen education turn into a business.

In the United Kingdom, during the era of Margaret Thatcher (former British Prime Minister), much more emphasis was put on higher education. Technical colleges were converted into universities. Admission criteria were changed. She introduced universities to market forces, by introducing fees for international students. International students have soared ever since. At the same time the government needed to sell universities to British teenagers and their parents. After many years, the process has been mis-sold. In the new world of tuition fees and debt, children were led to believe that degrees earn big salaries. Students are now coming out of the universities with debt rising to £50,000, owing to the student loan company. Unfortunately, the harsh reality is that coffee shops and call centres become strewn[199] with university graduates.

How can we convince the government that there is no point in creating more graduates unless you have more graduate-level jobs?

Research published by Oxford University economists Ken Mayhew and Craig Holmes found that "the UK now has proportionally more graduates than any other rich country bar Iceland" –

yet these graduates then tend not to use the academic skill that they have gained when they enter the working world (Holmes & Mayhew, 2015). Graduates are now doing jobs that previously didn't require a degree. One in six members of staff at call centres have a university degree. It is the devaluation of the academic degree that is the crux of the issue.

11.3 EDUCATION AND MORALITY

Nelson Mandela famously said that education is the most powerful weapon that can be used the change the world. Change through education is at the core of reaching for higher initiative. Reaching a higher level of education not only arms young people with the knowledge they need, but also actively engages them in important aims for their lives. It helps them to amplify their ambitions and dreams.

However, methods of education have undergone constant transformation since ancient times. Over the centuries we have seen great refinements in civilizations' attitudes and social developments. There have been changes in the understanding of nature and the universe. Education has uprooted many superstitions and social maladies[286]. Many social evils have been eradicated[74] through education. Along with the revolutionary changes made by human knowledge, we also find 'moral degeneration[287],' and a rapid decline in old values.

The modern education system is producing self-centred individuals who have little care about society in general. To them, it is all about, me, myself and I! The result is lack of respect, sympathy and empathy! Unfortunately, 'grades and scores' have become the sole reason for modern education.

Some scientific studies assert that valid knowledge can only be achieved from sciences. They hold the view that religious beliefs are the remains of pre-scientific explanations of the world and are nothing but superstition. Some people of faith believe that science only conveys a materialistic view of the world. However, it is faith that at the base of things, made possible the growth of science.

Scientific and religious beliefs need not be in any contradiction if they are properly understood. The scope of science is in the world of nature, mathematically observed and based upon changeable theories. Outside of the world of nature, science has no authority, no statement to make! Science has nothing to say, decisively, about values, morals and the meaning of life. It has nothing to say about religious beliefs, except when these beliefs make assertions about the natural world that contradicts scientific knowledge.

Whether there is a purpose to the universe or a purpose for human existence, these are not the questions for science. We must hold strong beliefs about the moral values that religions have given us over thousands of years. This should work with the acceptance of the occurrence of 'evolution[118]'.

The technology derived from scientific knowledge pervades[141] our lives. High-rise buildings of our cities, long-span bridges, rockets that take us into space, telephones and computers that perform complex calculations in seconds, drugs and gene therapies that replaces DNA in defective cells are some remarkable achievements of scientific knowledge.

However, individuals were supposed to direct their personal behaviour towards the salvation of their soul and their political behaviour around the state. Morality is not declining in the modern world, instead, a new morality is emerging. This morality is centred on 'self-fulfilment' and is linked to state-controlled values. These values are portrayed by the mass and popular media, which in turn is virtually controlled by the establishment.

Sometimes I wonder: who is a bigger offender, masters or fundamentalists? The UN has become a toothless tiger since it spends billions of dollars as a peacekeeping force rather than a peace preventative body. We also have to find ways of promoting a peaceful coexistence with religious tolerance through literacy and basic education. I advocate the importance of education as a forceful tool for the achievement of social and moral values. There must be an international standard classification of education, at least at the primary school level. It should be a priority for developing countries.

CONCLUSION

In my study, I have tried to unravel 'taboos' attached to the creator of the universe, evolutionary theory and an account of humans, and their morality. I have also analysed a historian's view of the origin of civilization and religion. In 'Social Aspects', I have addressed culture and society as well as state formation and their controlling tools: banking and media.

I have used science to describe the creation of the universe. Science is my way of understanding God. Through this it is much easier to see how everything works in the universe with perfect harmony. This, to my mind, is substantial proof of a creator that has designed the universe with empirical[111] order.

Perhaps God and science are not opposite. Biology does not undermine God, and one can argue that it merely illustrates God's creative power. Perhaps God implanted within nature a way to be evolved. I believe that God and Science are not at odds. They reveal each other's powers. Of course, some scientists will argue against this view. On the other hand, if we proclaim that God created the universe. Then who created God? So, what happened before the Big Bang? The line of questioning becomes uncomfortably difficult. Maybe time, space, and cosmic set up all began with The Big Bang. There was no 'before' in which the great cosmic drama occurs. The burst of energy and light was the singular start of everything.

The earth exists with a plethora[288] of finite[69] physical conditions that enable our existence. We live on a planet that has all the correct ingredients for life. The earth orbits the sun at a specific distance such that the planet temperature is conducive[289] to life. The correct amount of liquid water, and an atmospheric pressure that facilitates the existence of this liquid water at our surface enables a world with both oceans and continents. Earth, to our current knowledge, is the only planet with the correct composition of gases to sustain plants, animals and human life.

Even though scientists are establishing their theories based on empirical[111] calculations whilst they progressively explore the universe, the human mind is programmed to believe that our existence cannot be just cold, hard numbers. 'Faith' adds a notion of 'morality' and 'ethics' into our existence. I believe that there is nothing wrong in having faith and spirituality, which gives one comfort and hope of good things to come.

All great religions have their account of the evolution[118] of life in our universe. Scientific theories of the evolution of humans are not a million miles away from progressive creationism theories, which are more favoured by biblical text. However, instead of involving myself in a pointless discussion, I am more interested in how the human community developed in stages to from group living, and how morality evolved and grew out of that primitive society.

Finally, I have touched upon the socio-cultural evolution[118] of family, marriage and education. For each topic, I have investigated its historic evolution and how it has developed up until today's world! I have discovered that at present, morality is chosen by the people as for the purposes of convenience. It is filled with an 'anything goes' mentality. Truth is blurred with lies, hypocrisy and misinformation. Consequently, we are creating an illusion of reality separate from the truth. Reality is not a notional idea of how the things exist; it should be the innermost essence of all that exists.

Through this journey of discovery that I embarked on, I am fascinated that virtually every society around the world had established the values of 'Good and Evil'. These societies may be living in different parts of the world in different centuries, completely isolated from each other. Yet they managed to establish the same parameters of good and evil as highlighted in 'Good and Evil'.

It is clear in the contemporary world that society is using the media to promote its own values and views. Today we are in a role-reversal society. More children are defining the boundaries rather than their parents teaching them core values. There is little or no discipline for poor behaviour, people no longer care about how their interactions with others affect the outside world. Today, everything is about instant pleasure and gratification. Anything that gives a shot of pleasure is acceptable. If you wanted to, you could literally spend the whole day looking at pictures of yourself and doing nothing but talking about yourself. None of your thousands of Facebook friends will say anything, as this in itself would be hypocritical of them.

This generation knows not what respect is; they don't respect themselves, or others. The majority are not aware of what their morals and values should be. We now live in a society where we see kids disrespecting elders and talking down to them. This would have resulted in a slap across my face, back in my day. Teenagers are guided by lust instead of having self-control. The brand of your shirt or shoes is more important than the grades you achieve in school. A decline in respect for school teachers and the people of authority is eroding[290] our society.

Every generation feels that the next generation is changing and becoming worse. This could simply be social evolution[118] in progress. The question is, what do we value now? Considered virtues such as respect, empathy and honesty are quickly disappearing and being replaced with material things and instant gratification.

As discussed above, the atrocities that children have been subjected to in the past cannot be understated. This was just as immoral as the attitude of the children themselves today. The real challenge for our future is to find a balance that allows children to flourish creatively, but at the same time, instil the moral values that are lacking in today's youth.

Different cultures, religions, philosophies and organisations have varying concepts on how to achieve peace in the world. They are all aiming to achieve this by addressing human rights, education and diplomacy. Since 1945, the United Nations and the permanent members of its security council have endeavoured to resolve conflicts without war. Nevertheless, nations have still entered into military conflicts. It is a particular theory relating to US forces combined with the allies trying to attain peace through force and strength, that has failed miserably. The reason behind this failure is that the US has a personal agenda and a lack of policy which should treat all parties involved in conflict with neutrality and justice.

The traditional diplomatic strategies of influence such as economic sanctions, withdrawal of foreign aid and threatening to use direct military force have failed. President Trump has drawn praise from Israel for withdrawing from Iran's nuclear deal and for making a point of moving the US embassy to Jerusalem. The move has been perceived as taking sides in the disagreements between Israel and Palestine. Therefore, it is no longer accurate to describe the United States as neutral arbiters[291] in the dispute. The United States National Security Advisor, Mr. John Bolton, speaking in 2000 stated:

"If I were redoing the UN Security Council today, I would have one permanent member, because that is the real reflection of the distribution of power in the world" (*The New York Times*, 2005).

This **reality**, whereby an ignorant and dismissive American voice claims to be the best peace broker between two independent nations, is what lies at the core of the problem. US policy is no longer 'America First' but rather 'America Alone'. As for the **morality** of this issue, disillusionment and a lack of input from the rest of the world in key decision making is becoming a major problem.

GLOSSARY OF KEY WORDS

Ref. N	Word	Urdu Translation
1	Contrived	ایجاد کرنا، نقلی، بناؤٹی، مصنوعی
2	Preconception	عصبیت، پیش بینی، تصوّر
3	Synthesised	ترکیب دینا، کلیہ بنانا
4	Validity	دُرُست
5	Enmity	دُشمنی
6	Antagonism	حریفانہ، مخالفت
7	Entity	وجود
8	Retribution	سزا، عزاب
9	Forged	جعلی
10	Irreprehensible	بے عیب
11	Ontological	وجودیاتی
12	Promulgation	نفاذ، اعلان کرنا
13	Propensity	رجحان، رغبت، خواہش
14	Obligatory	لازم، ضروری
15	Metaphysical	روحانی، غیر مادی
16	Elucidate	واضع کرنا

17	Dogmas	ہٹ دھرمی
18	Book of Genesis	تورات کی پہلی کتاب
19	Embryology	ماں کے رحم میں پرورش اور نشونما سے مطالق علوم
20	Assigned	سونپنا، حوالے کرنا
21	Autonomous	خود مختار
22	Realm	دائرہ کار
23	Natural Phenomenon	قدرتی رجحان
24	Provisory	مشروط
25	Derivation	ماخذ، اخذ کرنا
26	Prevailing	مروجہ، غالب ہونا
27	Annihilation	فنا، مٹا دینا
28	Correlation	باہمی تعلق
29	Metaphor	استعارہ، مجازی
30	Nuclei	مرکز، کلیدی حصّہ
31	Isotopes	تمام اطراف اور خواص میں یکساں
32	Cosmic Microwaves	خلا و کہکشاں سے گزرتی کائنات پر پڑنے والی شعائیں
33	Quantified Manifestation	خاص مقدار میں ظہور پذی
34	Eludes	آنکھ سے اوجھل
35	Genetic Code	خلقی، وراثتی ظابطہِ اخلاق

36	Respectively	بالترتیب
37	Monumental	یادگار
38	Transcendent	اعلیٰ، بالاتر
39	Omnipresent	ہر سُو، ہر جگہ موجود
40	Omnipotent	قادرِ مُطلق
41	Infinite	لامحدود
42	Eternal	ابدی
43	Cosmos	عالمِ ظاہر، کائنات
44	Biogenesis	نظریہِ حیات
45	Creationist	عقیدہِ تخلیق، تخلیق کرنے والا
46	Infallible	یقینی، غلطی سے مبّرا
47	*Homo sapiens*	نوعِ انسانی
48	Neanderthal	گنوار آدمی، جنگلی
49	*Homo erectus*	انسان نما مخلوق
50	Fossil	زمین کھود کر نکالا ہوا قدیم عنصر، جز
51	Interbred	دو نسلوں کو ملانا
52	Cashing	داغدار
53	Caucasian	سفید فام باشندے
54	Circumventing	پھیل جانا، جل دینا
55	Equator	خطِ استواء
56	Indigenous	مقامی، اصل باشندے
57	Meridian	عروج، بلند
58	Oceania	بحرالکاہل اور قریبی

		سمندروں کے جزیروں کا مجموعہ
59	Solidarity	یکجہتی
60	Artisans	ہنرمند
61	Primate	امام، اعلیٰ مخلوق
62	Shamans	سائبریا کے مزہبی عقیدے کے مطابق وہ شخص جو روحوں کو قابو کرے
63	Social Realm	سماجی دائرے
64	Adaptive Value	حسبِ منشا
65	Stratification	درجہ بندی
66	Pastoral	چرواہے
67	Perpetuating	ہمیشہ جاری رہے
68	Predatory	غارت گری، وحشیانہ، لوٹ مار
69	Finite	محدود
70	Eternity	ہمیشگی
71	Clandestinely	پُراسرار،چوری چھپے
72	Enactment	فرمان
73	Obliterate	منسوخ کرنا، ختم کرنا
74	Eradicate	مٹانا
75	Omniscient	علم الغیب، ہر چیز کو جاننے کی صلاحیت
76	Omnibenevolent	کریم مطلق، کرم کرنے والا

77	Supernatural	مافوق الفطرت
78	Aspires	خواہش مند
79	Frugality	کفایت شعاری
80	Antithesis	برعکس
81	Innate	فطری
82	Enlightened	روشن خیال
83	Cognitive	علم رکھنے والا، ادراکی
84	Comprehension	سمجھ
85	Non-rational	غیر عقلی
86	Doctrine	نظریہ
87	Fideism	عقیدہ
88	Elevated	بلند
89	Intuition	الہام
90	Conform	مطابق
91	Paradox	خلافِ قیاس
92	Transcendence	فوقیت
93	Humility	عاجزی
94	Sermon	خطبہ
95	Incorporated	شامل کرنا
96	Compassion	ہمدردی
97	Naturalistic	فطرتی، مادہ پرستی
98	Elaborate	تفصیل، وضاحت
99	Apocalypse	الہامی کتاب

100	Complementary	اعزازی، امدادی
101	Abyss	بے اندازہ، گہرائی
102	Barzakh	عالمِ ارواح
103	Cessation	خاتمہ
104	Organism	حیاتیات، ہستی
105	Divinity	معبود
106	Dichotomy	دو اجزاء میں تقسیم
107	Commandments	حکمِ الٰہی
108	Resurrection	دوبارہ زندہ ہونا
109	Skepticism	نظریاتی شک
110	Quantum Physics	نظریہ مقدار، برقیات
111	Empirical	تجرباتی، عملی مشاہدہ
112	Variable	تغیر پذیر، بدلنے والا
113	Hypnosis	مصنوعی نیند
114	Holotropic	سانس لینے کا منفرد طریقہ
115	Spheres	دائرہ نما، مدار صفت ، گول شکل
116	Transgression	حکم عدولی
117	Harassment	ہراساں کرنا
118	Evolution	ارتقاء
119	Hominid	نوعِ انسانی
120	Neocortex	دماغ کی بیرونی تہہ
121	Dopamine	دماغ میں موجود مرکب

122	Hierarchy	تنظیمی نظام، درجہ بندی
123	Funerary	کفن دفن
124	Deities	دیوتا
125	Sages	عالِم، دانا، عاقل
126	Polytheistic	مشرکانہ
127	Paganism	مظاہر پرستی
128	Unperturbed	غیر مضطرب، مطمئین
129	Transcendental	بالاتر
130	Conversely	اس کے بر عکس
131	Pantheon	تمام دیوتاؤں کا مندر
132	Avesta	پارسی کی مقدس کتاب
133	Zoroaster	آتش پرست
134	Proto-Indo-European	آرائین
135	Monotheism	عقیدہِ توحید
136	Culmination	نقطہِ عروج
137	Analogical	مطابقت
138	Diaspora	تارکینِ وطن
139	Anthropology	علم بشریات، علم الانسان
140	Conceptually	خیالی طور پر
141	Pervades	سما جانا
142	Reincarnation	تناسخ، روح کا ایک قالب سے دوسرے

		قالب میں جانا
143	Manifestation	اعلان، مظہر
144	Disciple	مرید، شاگرد
145	Ambiguous	مبہم، مشکوک
146	Aversion	نفرت
147	Sublime	شاندار
148	Austere	سادگی
149	Celibate	کنوارا پن
150	Denigrate	بدنام کرنا
151	Licentious	آواره
152	Communal	فرقہ وارانہ
153	Infer	قیاس کرنا
154	Culminate	اونچائی پر لے جانا
155	Covenant	عہد
156	Ad hoc	وقتی
157	Abomination	مکروہ، نفرت انگیز
158	Persecuted	ایذا رسائی، پیچھے پڑنا
159	Adherent	پیروکار
160	Apostolic	منصبِ رسالت
161	Epistle	مراسلہ
162	Purgatory	وہ مقام جہاں روحوں کو پاک کیا جاتا ہے
163	Purged	کفارہ
164	Diocese	ڈیکن (پادری) کے

		ماتحت علاقہ
165	Sacraments	مقدّس رسوم
166	Seminarian	مزہبی درسگاہ
167	Excommunication	دین بدر
168	Schism	فرقہ بندی
169	Synod	ارکانِ کلیسہ
170	Animistic Polytheism	مظاہر پرستی، بت پرستی
171	Allegiance	بیعت، اطاعت
172	Depicts	دکھایا جانا
173	Heterogeneous	مختلف نسل، غیر جنس
174	Sassanid	ساسانی، ایرانی
175	Autocratic	مطلق العنان، بے لگام، خود سر
176	Instilled	آہستہ آہستہ، قطرہ قطرہ
177	Austerity	سادگی
178	Piet	تقویٰ
179	De facto	حقیقی قبضہ
180	Repression	جبر
181	Hereditary	موروثی
182	Faction	گروہ
183	Schismatic	بدعتی
184	Fervor	گرمجوشی
185	Apogee	بلند ترین مقام

186	Prevalent	مقبول
187	Veered	رُخ بدلنا
188	Stifled	گلا گھونٹنا
189	Egalitarian	عقیدہِ مساوات
190	Polygamy	ایک سے زیادہ شادیاں
191	Patriarchal	سرداری
192	Susceptible	اثر پزیر
193	Diacritical	امتیازی
194	Allegorical	مثالی، تمثیلی
195	Perplexed	الجھن کا شکار
196	Fundamentalism	بنیاد پرستی
197	Resurgence	حیاتِ نو
198	Genocide	نسل کشی
199	Strewn	بکھرے ہوئے
200	Authentication	تصدیق
201	Succumb	مغلوب
202	Subjugation	محکام،تابعدار
203	Solace	تسّلی، دلاسا
204	Raw Fodder	چارہ ڈالنا
205	Frenzy	جنوں
206	Delve	تحقیق
207	Dynamics	متحرک
208	Predisposed	راغب، مائل
209	Trait	خاصیت

210	Vocation	پیشہ
211	Sitcom	مزاہیہ
212	Perverse	گمراہ
213	Creed	عقیدہ
214	Subverted	تہہ و بالا کرنا
215	Concomitantly	لازم و ملزوم
216	Kinship	رشتے داری
217	Rearing	پرورش کرنا
218	Proliferated	وسیع ہونا
219	Pederasty	اغلام، ہم جنس پرستی
220	Conceptualised	تصّوراتی
221	Unilateral	یکطرفہ
222	Punitive	تعزیری، عقوبتی، سزا
223	Concubine	لونڈی
224	Celibacy	کنوارہ پن
225	Coverture	حالتِ ازدواج
226	Malicious	بد نیتی
227	Cyberbullying	خیالی غنڈہ گردی
228	Fragmentary	ادھوری
229	Rhetorical	بیان بازی
230	Harnessed	فائدہ مند
231	Piercing	چبھنے والی
232	Contrition	شرم ساری
233	Culprits	مجرم

234	Ascendancy	عروج
235	Slander	تہمت لگانا
236	Pedantic	نمائشی
237	Array	صف آرائی
238	Hypothesis	مفروضہ
239	Arid	بنجر
240	Circumscribed	احاطہ شدہ، محیط
241	Segmentation	علیحدگی کا عمل
242	Genesis	ابتداء
243	Dissolution	تحلیل کرنا
244	Fledgling	نو خیز
245	Dogma	اصول
246	Veto	نا منظوری
247	Violation	خلاف ورزی
248	Barter	لین دین
249	Levied	لگان
250	Constraints	رکاوٹیں
251	Renaissance	علمِ فنون کے احیاء کا دور
252	Provoked	اکسانا
253	Usury	بھاری سود خوری
254	Gentiles	غیر یہودی
255	Deuteronomy	تورات کی پانچویں کتاب

256	Victuals	خوراک
257	Riba	سود
258	Promissory	اقراری یادداشت
259	Bond	معاہدہ
260	Inflation	افراطِ زر
261	Solicits	التجا کرنا
262	Extorts	دھمکی دینا
263	Pervasive	سرائیت کرنا
264	Niche	مقامِ متعین، مناسب جگہ
265	Synergy	متحدہ عمل
266	Disburse	ادائیگی
267	Repetitive	بار بار دھرانے والا
268	Synonymous	مترادف
269	Lyceum	درسگاہ
270	Fostered	فروغ دینا
271	Pyramid	اہرام
272	Sphinx	ابوالہول، قدیم مصری مجسمہ
273	Plateau	سطح مرتفع، بلند ہموار مقام
274	Odyssey	تجرباتی سفر
275	Monasteries	خانقاہیں
276	Mediate	ثالثی
277	Compulsion	مجبوری

278	Qualms	دِقت
279	Bourgeois	ادنیٰ درجے کے خیالات والا انسان
280	Reformation	ازسرِنو تشکیل دینا
281	Automated	خودکار مشین
282	Cusp	چوٹی، بلندی
283	Sadistic	ایذا پسندی
284	Intrinsic	پیدائشی
285	Indictive	جتانے والا
286	Maladies	غلط روّیے
287	Degeneration	ذوال
288	Plethora	ڈھیر سارے
289	Conducive	سازگار
290	Eroding	تباہ کرنا
291	Arbiters	ثالث

INDEX

9/11:	On Sept 11 2001 American Airlines Boeing 767 loaded with 20,000 gallons of jet fuel crashed into the north tower of World Trade Center in New York City.
Abiogenesis:	Original evolution of living organisms
Abraham:	Father of nations, played significant part in Judaism, Christianity and Islam.
Age UK:	British women having children later than in any other country
Al Burani:	Great Persian scientist who measured the diameter of the earth (973-1048)
Alexander the Great:	King of Macedon, conquered Anatolia, Persia, Egypt, Asia Minor and Bactria (356-323 BC)
Al-Ghazali:	(1058-1111)
Al-Hallaj:	Mansur Al Hallaj (858-922) Persian poet, teacher of Sufism who claimed (Ana I-Haqq) I am the truth. He was martyred.
Apocalypse:	Disclosure of knowledge which is hidden.
Apostles:	Twelve disciples of Jesus Christ.
Aristotle, Plato, Socrates:	Greek philosophers.

Bertrand Russell:	British philosopher, historian and writer: en.wikipedia.org/wiki/Bertrand-Russell
Centre of Disease Control and Prevention USA:	CDC is a federal agency that conducts and supports health promotion.
Charlemagne:	Former King of Franks known as Charles the Great. He united much of Europe during Middle Ages (742-814)
Christopher J. Ferguson:	An American psychologist who serves as a professor at Stetson University, Florida.
Constantine the Great:	Built Constantinople, present-day Istanbul.
Copying:	Cultural transmission depends on copying, understanding cultural transmission in anthology.
Covertures:	Legal doctrine in USA, which dissolves bride's identity to confirm husband's dominance.
Cyrus the Great:	The founder of the first Persian Empire.
David Spergel:	American theoretical astrophysicist at Princeton University (1961-) Prize-winning cosmologist.
Deoband school:	(Darul uloom) Islamic school in India's Saharanpur district. It started Islamic revivalist and anti-imperialist movement.

Edwin Hubble:	(1889-1953) American astronomer, played a crucial role in observational cosmology.
Einstein:	German scientist who discovered the Theory of Relativity ($E=mc^2$). Nobel Prize winner in physics (1879-1955)
Elements of culture:	Eight basic elements of culture.
Engels:	Friedrich Engels, a German philosopher and social scientist.
European Renaissance:	An extension of Middle Ages from 14th-century European history.
Exile:	Period of history during which a number of people from Juda were captives in Babylon.
First temple:	Solomon's temple in Hebrew (Belt Ha Mikdash) destroyed in 587BC
Francisco Pena:	Jose Francisco Pena Gomes, politician from the Dominican Republic, leader of the Revolutionary Party, died 19th May 1998.
Franz de Waal:	Dutel Primatologists. Professor of Primate behaviour (1948-)
Genesis:	The first book of Hebrew Bible.
Hazrat Ali:	(599-661) Fourth and final of the Rashidun Caliphs.
Hellenistic culture:	Period between Alexander and Romans.
Higgs boson:	Professor Higgs (Nobel Prize winner), theoretical physics, discovered God Particle, 1929.

Homeric gods:	The gods in Iliad (the account of Trojan War and odyssey)
Homo erectus:	Extinct species of humans. Fossil evidence dates to 1.9 million years ago.
Homo sapiens:	Systematic name used for anatomically modern humans
Immanuel Kant:	German philosopher in ethics and metaphysics (1724-1804).
Indian narrations of Epics V:	Qur'an theology.
Islamophobia:	Dislike of or prejudice against Islam or Muslims especially as a political force.
Johann Hamman:	German philosopher (1730-1788)
Ken Mayhew and Craig Holmes:	Ken Mayhew is a professor of education and economics performance at Pembroke College, Oxford. Dr Craig Holmes' research interests are in labour, earning equality and skill.
Kharaj and Jizya:	Kharaj, the land tax to be paid by landholders, and Jizya by non-Muslims.
Leonardo Da Vinci:	Italian painter, architect and inventor.
Lord Sacks:	Jonathan Henry Sacks, Baron Sacks MBE. British orthodox rabbi, philosopher and author.
Margaret	British Prime Minister from (1979-1990). She had immense impact on

Thatcher:	higher education.
Maryanne Wolf:	Educator and author, Director of centre for reading and language research.
Matt J Rossano:	Professor of evolutionary psychology at Louisiana University.
Naeem Tarar,	Edhi foundation.
Neanderthal:	Human, became extinct about 40,000 years ago.
Old Testament:	First part of Christian Bible based primarily upon Hebrew Bible. It contains Genesis, Exodus, Leviticus, Numbers and Deuteronomy.
PDPA:	People's Democratic Party of Afghanistan. Established 1st Jan 1965. Dissolved in 1992. Ideology: Communism.
Purgatory:	After death, the suspended stage of the soul. According to Catholic belief.
Robert Crookall:	(1890-1981) geologist, spiritual and consciousness studies.
Sharia law:	Islamic law. Rules which are part of the Islamic culture.
Statue of Zeus:	By the Greek sculptor Phidias in 435 BC. Considered a wonder of the world.
Syed Qurb:	Egyptian scholar, poet, theorist (1906-1966)
Talmud:	The central text of Rabbinic Judaism.
Tom Burgis:	Tom Burgis covered Africa for *Financial Times* for six years.

Tora (Deuternomy) 6:4	"Hear O People, the lord our God The Lord is one."
Victor Zammit:	Barrister, psychologist, writer and researcher on empirical evidence for 'after life'.

REFERENCES

Age UK, 2017. [Online] Available at:
https://www.ageuk.org.uk/
[Accessed 2017].

Ali, H., 10th Century. *Nahj al-Balagha.* s.l.:s.n.

ATLAS Collaboration, A. G. A. T. et al., 2012.
Observation of a new particle in the search for the
Standard Model Higgs boson with the ATLAS
detector at the LHC. *Physics Letters B,* 716(1).

Born, M., 2012. *Einstein's Theory of Relativity.*
s.l.:s.n.

Burgis, T., 2015. *The Looting Machine.* s.l.:s.n.

Christianson, G. E., 1996. *Edwin Hubble.* s.l.:s.n.

Crookall, R., 1969. *The interpretation of cosmic
and mystical experiences.* s.l.:s.n.

De Waal, F., 2016. *Are We Smart Enough to Know
How Smart Animals Are.* s.l.:s.n.

Demause, L., 1988. *History of Childhood: The
Untold Story of Child Abuse.* s.l.:s.n.

Encyclopaedia Britannica, 2018. *Babylonian Exile.*
[Online]
Available at:
https://www.britannica.com/event/Babylonian-Exile
[Accessed 2017].

Encyclopaedia Britannica, 2018. *Neoevolutionism.*

[Online]
Available at:
https://www.britannica.com/topic/neoevolutionism
[Accessed 2018].

Engels, F., 1884. *Origin of the Family, Private Property, and the State.* [Online]
Available at:
https://www.marxists.org/archive/marx/works/down load/pdf/origin_family.pdf
[Accessed 2017].

Furedi, F., 2015. *How the Internet and social media are changing culture.* [Online] Available at:
http://www.frankfuredi.com/article/how_the_intern et_and_social_media_are_changing_culture1
[Accessed 2017].

Global Resources, 2018. [Online] Available at:
http://www.global-resources.co.uk/ [Accessed 2017].

Gottlieb, A., 1997. *The Great Philosophers: Socrates.* s.l.:s.n.

Harari, Y. N., 2014. *Sapiens: A Brief History of Humankind.* s.l.:s.n.

Holmes, C. & Mayhew, K., 2015. HAVE UK EARNINGS DISTRIBUTIONS POLARISED?. *Institute for New Economic Thinking.*

Jackson, S. A., 2002. *On the Boundaries of Theological Tolerance in Islam: Abu Hamid al Ghazali's Faysal al Tafriqa (Studies in Islamic Philosphy).* s.l.:s.n.

Josephus, F., 93-94 AD. *Antiquities of the Jews.* s.l.:s.n.

Merton, R. K., 1949. *Social Theory and Social Structure.* s.l.:s.n.

NASA Science, 2018. *Dark Energy, Dark Matter.* [Online] Available at: https://science.nasa.gov/astrophysics/focus-areas/what-is-dark-energy [Accessed 2018].

Philosophy, S. E. O., 2010. *Immanual Kant.* s.l.:s.n.

Qutb, S., 1964. *Milestones.* s.l.:s.n.

Randall, J., 2009. No respect, no morals, no trust - welcome to modern Britain. *The Telegraph*, 5 November.

Rossano, M. J., 2010. *Supernatural Selection: How Religion Evolved..* s.l.:s.n.

Russell, B., 1945. *History of Western Philosophy.* s.l.:s.n.

Smith, H., 1958. *The World's Religions: Our Great Wisdom Traditions.* s.l.:s.n.

Spergel, D. N., 2015. The dark side of cosmology: Dark matter and dark energy. *Science,* 347(6226), pp. 1100-1102.

The New York Times, 2005. *Opinion- The World According to Bolton.* [Online] Available at: https://www.nytimes.com/2005/03/09/opinion/the-world-according-to-bolton.html [Accessed 2017].

The Telegraph, 2016. *World Peace Day: 17 inspirational quotes about peace.* [Online] Available at: https://www.telegraph.co.uk/news/0/world-peace-day-17-inspirational-quotes-about-peace/ [Accessed

2017].

Varner, W. & Varner, W. C., 1987. *Jacob's Dozen: A Prophetic Look at the Tribes of Israel.* s.l.:s.n.

Zanmmit, V., 2000. *Victor Zammit.* [Online] Available at: www.victorzammit.com [Accessed 2017].

ABOUT THE AUTHOR

Farooq Tareen was born in 1944 in India. He was four years old when his parents moved to Pakistan after the Indian subcontinent partition. The experiences of that horrendous journey were engraved into his mind and have left their ever-lasting mark. As he was growing up in Pakistan, he witnessed the resentment among people towards Western policies in the wake of the catastrophic expulsion of Palestinians in 1948. As he was growing up, he was influenced by social preconceptions of religion and through the accessible literature, making him struggle to find the correct worldview. In 2010, a life-changing event in which he was told he only had five months to live propelled him to investigate all about life, religion and the universe.